Sure Signs That You Need to Repair Your Plumbing

- ➤ Your pipes play the national anthem every time you turn on the shower.
- ➤ The water coming out of the tap is the same shade of green as your kitchen sink.
- ➤ The original plumber wrote his name and the year 1867 on the inside of your toilet tank.
- ➤ Your local museum of science and industry wants to place your plumbing system on the register of historic places.
- ➤ More water leaks out of your pipes than comes out of the taps.
- ➤ By the time your bathtub fills for your evening bath, it's time to eat breakfast.

Ten Common Plumbing Mistakes

1. Running pipes without insulation through unheated crawl spaces.
2. Ignoring small drips and leaks before they become bigger problems.
3. Not installing shutoff valves at every fixture.
4. Neglecting to take out a permit for any plumbing work that requires one.
5. Buying the wrong parts for faucet repairs.
6. Forgetting to partially drain your water heater in order to get rid of any sediment in the tank.
7. Putting off regular maintenance and cleaning of septic tanks.
8. Buying a faucet that doesn't match up with your sink.
9. Poor planning of pipe and vent runs.
10. [...]ining fixtures, a sure sign of a clog in progress.

alpha
books

Remodeling Money Savers

- ➤ Buy quality fixtures, but avoid paying for features you don't really need.
- ➤ Consider installing an acrylic tub/shower combination instead of installing tile walls.
- ➤ Try and add a new bathroom near an existing one so you can take advantage of the water supply and drains.
- ➤ Get at least three bids from qualified contractors for any major job.
- ➤ Carefully determine the best size of water heater for your household needs rather than buy a larger model than is suitable.

Jobs That Are Worth Hiring Out

- ➤ Water heater replacement
- ➤ Gas pipe work
- ➤ Replacement of all your pipes
- ➤ Installation of the main water supply pipe
- ➤ Steel or iron pipe replacement

Jobs You Can Do Yourself

- ➤ Replacing faucet washers
- ➤ Unclogging most stopped up toilets and drain lines
- ➤ Repairing minor pipe leaks
- ➤ Adjusting the water level in a toilet tank
- ➤ Flushing and draining a water heater

When You Need a Plumber

- ➤ Be clear about the work you want done or the repairs the plumber is proposing to do.
- ➤ Get an estimate up front before the work begins.
- ➤ Find out who will be doing the work—a journeyman plumber or an apprentice.
- ➤ Review the completed job with the plumber and check that everything is working satisfactorily.
- ➤ Pay promptly.

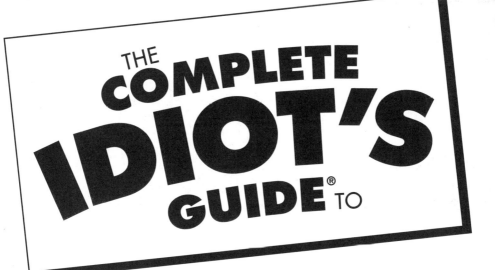

THE
COMPLETE
IDIOT'S
GUIDE® TO

Plumbing

by Terry Meany

alpha
books

Macmillan USA, Inc.
201 West 103rd Street
Indianapolis, IN 46290

A Pearson Education Company

Publisher
Marie Butler-Knight

Product Manager
Phil Kitchel

Managing Editor
Cari Luna

Acquisitions Editor
Randy Ladenheim-Gil

Development Editor
Joan D. Paterson

Production Editor
Billy Fields

Copy Editor
Krista Hansing

Illustrator
Jody P. Schaeffer

Cover Designers
Mike Freeland
Kevin Spear

Book Designers
Scott Cook and Amy Adams of DesignLab

Indexer
Angie Bess

Layout/Proofreading
Angela Calvert
Mary Hunt
Julie Swenson

Contents at a Glance

Contents

Appendixes

Foreword

In the early 60s, I remember my father bringing home coffee mugs from the plumbing wholesaler. They were large white mugs with a soft-blue image of a man carrying a pipe wrench. Next to the image were the words "Plumbers protect the health of the nation." Historians contend that the Bubonic Plague, which raged throughout Europe, resulted from improper sanitation. Today, whenever a disaster strikes a community—whether flood, tornado, or earthquake—the first step is to provide safe water for drinking, cooking, and medical procedures. Environmentalists are always talking and screaming about water pollution, not only of our surface waters, but also of our subterranean water sources. Thus, we see that plumbers do protect the health of the nation.

Each and every plumbing installer has a responsibility to perform plumbing installations that ensure health safety. Not only do the completed jobs need to be properly installed, but their maintenance is critical to the well-being of the direct consumer as well as to the health of the community. In addition, the plumbing installer must perform the task so as not to create any endangering situations to the technician or the structure.

The Complete Idiot's Guide to Plumbing explains in clear and simple terms, both pictorially and in writing, the basic requirements for any lay person to perform plumbing assignments that will produce positive results in an efficient and safe manner. This book provides an accurate map to maintain and install the numerous components of a plumbing system: fixtures, water heaters, traps, water piping, gas piping, and sanitary drain and vent piping. Anyone relying on this book will be able to produce an installation and come up with a sound maintenance program that will provide years of reliable service while protecting the health of homeowner, the community, and the nation.

Harvey Kreitenberg

Mr. Kreitenberg is an expert witness for construction defects in the world of plumbing. By following *The Complete Idiot's Guide to Plumbing,* an installer is unlikely to face Mr. Kreitenberg or ask him to defend any installation.

Introduction

Every home, from the humblest, one-bedroom cottage to the grandest megahouse, has a plumbing system. It could be argued that plumbing is our most critical home mechanical system because it actually keeps us alive by providing a daily supply of potable water. We can get along without electric lights and central heating, and even without cable TV (although some sports fans would disagree), but we can't survive without water. Aside from ensuring our survival, our plumbing system also helps keep us, our clothes and dishes, and even our cars clean. It's a marvel of simplicity and design, but sometimes we need to repair it, tweak it, and expand it.

Your one-bedroom family home of the 1920s might well be a cute, cozy bungalow with all its original charm. But there's nothing charming about 80-year-old pipes, and you don't have to live with them. Newer homes also can benefit from plumbing additions, such as an extra bathroom or a sprinkler system for the yard. And even if you're satisfied with your plumbing, you always face the potential of a leak or a clogged drain. Unlike self-cleaning ovens, our plumbing systems need some occasional TLC and that needs to come from either you or a plumber. You might not want to perform major repairs, but you should be knowledgeable enough to keep from calling a plum-ber for a simple washer replacement in a leaky faucet.

Don't know galvanized pipe from copper tubing or PVC from ABS? We'll walk you through your plumbing system, from the water meter to the main stack, so that you'll know how all those pipes and valves and fixtures go together. Plumbing systems are elegant in their simplicity and easy to understand—you could almost use them as an inspiration for your meditation sessions (well, you might want to leave out the toilets).

Plumbing work has one sword of Damocles that comes with it, however: Poorly done repairs will leak and possibly flood. You *can* get a second chance to do it right, but you'll have some water to clean up first. Follow the procedures outlined in this book, though, and you should stay high and dry. Your reading won't make you a journeyman plumber, but it will point you in the right direction.

As you read the following chapters, you'll learn more about your local water system and early American plumbing practices than you ever knew existed, and you'll also get a clear idea of what happens once the water enters your home. Then you can decide what else you want it to do and where you want it to go. Wherever you've got the room to run the pipes, you have a potential bathroom, laundry sink, or steam room.

Plumbing work often involves more than pipes and drains, too. You need to plan, budget, and take into account repairing walls and floors as you make access for the pipes. You'll see how to break large jobs into smaller tasks and stretch the job over a number of days, if necessary, without shutting off all the water to the rest of your house. Remember why you're doing this: to improve your home and make it more comfortable, not to involve yourself in an exercise in frustration.

How to Use This Book

There's an old Three Stooges short where Curly (of course) attempts to repair a leak, eventually enclosing himself in a jail of pipes and fittings with multiple leaks. I think this is what many people envision their plumbing repairs to be like. Fortunately, plumbing brings out the hesitation in all of us, and that puts the brakes on any rash moves with pipe wrenches and tube cutters.

This book is set up to first give you a broad overview of water systems, plumbing codes, how to hire a plumber, and how to do both minor and major repairs. You'll start with the small repairs and move up to more elaborate—and time-consuming—jobs. You'll decide how far you want to go before calling in a plumbing contractor.

The main rule you must follow, besides knowing when to jump in and when to pull back, is to adhere to your local plumbing code and have your work inspected. The rest is up to you.

How This Book Is Organized

Part 1: "The Basics: Civilization Equals Clean Water"

Knowing the importance of your fresh water system and how it comes into your home leads to a greater appreciation of it. Understanding how your own plumbing system works allows you to explore how you can improve or add to it, and assess whether the work falls under the do-it-yourself category or that of a plumber.

Part 2: "Tools, Techniques, and Fixtures"

The right tool for the job always makes that job easier to carry out and produces better results. Plumbing calls for its own tools, parts, and procedures, and you'll need to be familiar with them. Once the pipes are in, they have to end someplace, and that's usually at a fixture or a faucet. You'll be surprised at the selection you have to choose from.

Part 3: "Fundamental Fixes"

You have to start somewhere, and here you'll find the common drips, leaks, and clogs that most homeowners run into from time to time. Of course, an entire chapter is devoted to toilets!

Part 4: "Bigger Fixes"

By now, you might be ready for tougher challenges, and Part 4 won't disappoint. Still doable for a homeowner with time and perseverance, these jobs will definitely meet your criteria for a challenge.

Part 5: "Major Upgrades and Remodeling"

Here we move into the big leagues, everything from creating new bathrooms to replacing all your pipes. Not everyone wants to tackle these jobs, but you'll know what's involved and can decide on your own level of involvement and how much you want to hire out.

Part 6: "Other Plumbing Concerns"

This is the odds-and-ends section with chapters on hot water heaters, laundry rooms, and hot water and steam heating.

Extras

Some ideas need to stand out so that they don't get missed or to add interest to a subject. In the four types of boxes described here, I've highlighted information that provides easy problem solving, explains terminology, warns you of difficulties or dangers you might face, and offers bits of history and background on plumbing and plumbers.

Plumbing the Depths

This sidebar relates on-the-job experiences and intriguing plumbing trivia.

Pipe Dreams

Here you'll find tips to keep your home plumbing system in good shape.

Plumbing Perils

Pay attention to these boxes to keep plumbing disasters to a minimum.

What's That Thingamajig?

This box explains terms used in the plumbing industry and tells about the purpose of tools used in home plumbing repairs.

Acknowledgments

It might not take a village to write a book, but it takes collaborators, editors, and e-mail correspondents to pull it off. I'd like to thank Mark Riggs at Mansfield Plumbing Products, Inc.; Newbold Warden, Doug McLean, and Tammy Walz at TOTO; Paul DeBoo at Sloan Flushmate; and Maureen Namanich at Kohler for their help in supplying artwork and information on plumbing fixtures and faucets. Mike Mangan at MKM Communications, Joyce Simon at Western Forge, Raymond Venzon of Makita USA, and Sears Craftsman Tools all supplied tool art. My editors at Macmillan USA, Joan Paterson and Randy Ladenheim-Gil, shepherded the manuscript from e-mail files to hard copy. Terry Love of Love Plumbing & Remodel edited for the finer points of the plumbing trade. Lastly, I want to thank my agent, Andree Abecassis, for finding me such interesting work. I know more about the history of plumbing now than I thought existed.

Trademarks

Part 1
The Basics: Civilization Equals Clean Water

One of the great equalizers in life is something we never think about: potable water. Dr. Lewis Thomas recognized the historical importance of clean water and its role in our having gotten this far as a civilization. In the spring 1984 edition of Foreign Affairs, *the late Thomas said: "There is no question that our health has improved spectacularly in the past century. One thing seems certain: It did not happen because of medicine, or medical science, or even the presence of doctors. Much of the credit should go to the plumbers and (sanitary) engineers of the Western world."*

Thomas went on to note how waste-contaminated drinking water was once "the single greatest cause of human disease and death among us" and noted the virtual elimination of typhoid fever, cholera, and dysentery from the lives of Americans. The novelist John Gardener drove the point home as well when he said, "An excellent plumber is infinitely more admirable than an incompetent philosopher."

Your plumbing system consists of pipes, drain lines, valves, faucets, clean-outs, traps, vents, and more. Most of it is probably foreign to you, so this first part familiarizes you with the source of your water, how it gets to your house, and what happens once it's inside. You might be reluctant to add a bathroom on your own, but a little understanding of your system goes a long way, especially when troubleshooting problems. And you might have a newfound appreciation whenever you open a tap.

Chapter 1

Open Tap, Get Water

In This Chapter

➤ Understanding the water cycle

➤ Water treatment and sewage treatment

➤ Getting water to your house

➤ Moving water inside your house

➤ The importance of the drain-waste-vent system

Some of the greatest inventions and discoveries are the ones we take the most for granted. We turn a faucet handle and expect to get clean, safe water. Because we don't have to carry it inside by the bucket from a well in the backyard, we rarely think about the systems that bring in potable water and take waste water out of our homes. Sewage treatment plants aren't exactly the stuff of dinnertime conversations.

Before you declare that water works and waste plants could interest only water treatment engineers, consider your role. You're not only a consumer (unless you collect all your water in rain barrels), but you also affect the system. Everything you put down the drain goes through a sewage treatment plant. You'll have a much bigger effect on water quality if you decide to pour motor oil down your kitchen drain instead of olive oil.

During times of water shortages, rates might go up. Your demand for water—either intentional or due to leaks that go without repair—will affect your budget and could even instigate water restrictions on the part of your local utility if too many other households don't curtail their demands. Water shortages and excesses are all part of the water cycle. We're still subject to it, just as our ancient ancestors were—we just have better control over it.

What's That Thingamajig?

Potable refers to any liquid that is suitable for drinking. Suitable implies that the liquid has no harmful elements in it such as certain forms or amounts of bacteria. It does not imply that the liquid has an agreeable taste.

Water, Water, Everywhere

We're supposed to drink eight 8-ounce glasses of it every day. We wash in it, cook with it, and freeze it, and Cirque du Soleil clowns squirt in on us (tip: stay out of the front rows if you ever see their show). Water is the compound H_2O in liquid form. It's also ice when it's frozen, and steam when it's vaporized. It has no taste, covers the better part of the earth's surface, and even makes up the better part of us. We can live longer without food than without *potable* water.

Industrialized societies are the biggest users of water per person, in part because we have the infrastructure to collect, clean, and distribute it. Regardless of the sophistication of our technology, we have only so much influence over nature. Farmers understand this, but most of us city dwellers pay little attention to the hydrologic cycle.

Cycled and Recycled

The water cycle or hydrologic cycle is simply evaporation of water and its return to the earth in some form of precipitation. Water goes upward in the form of vapor and drops downward as rain or snow (all right, hail, too, for you meteorological purists). It then gets stored in bodies of water, in the ground, or as ice in such places as the North and South Poles. This cycle happens repeatedly, which means that the same water has been used for billions of years. This lends a whole new appreciation to sanitary engineers, considering where some of our drinking water has been.

As water passes through the ground, it picks up all sorts of mineral traveling companions, including these chemical concoctions:

➤ Sulfates

➤ Chlorides

➤ Oxides

Maybe it's time for a quick chemistry lesson. Minerals are inorganic (those without hydrocarbons) substances or compounds as opposed to those that are organic (containing carbon). In chemistry, a compound is a combination of elements and an element is a substance composed of similar atoms that cannot be separated by chemical interaction. Metals are elements, which allows me to segue ever so coolly into the subject of hard water. Hardness, in this case, simply means the concentration of magnesium and calcium ions present in the water. Both of these are metallic elements. Sometimes the presence of metals in water is confused with sediment. Small, loose particles of silt, clay, sand, and eroding soil are what make up sediment, any of which can be carried great distances by wind and water currents.

Plumbing the Depths

You would be hard-pressed to find pure, uncontaminated water anywhere in the world, even in the ice caps. Remote mountain streams serve as a water source for animals, who adulterate it with their presence or their wastes. Even ice core samples have been found to contain industrial pollutants. This has meant a sizable business for bottled water and water filtration systems, even though their health claims can be somewhat dubious. Purity itself is a relative term: What may be a passable level of water quality for agriculture, for instance, will most likely not be acceptable for drinking. The intended use of water will determine its quality.

Minerals are contained in the soil and rocks. Hard water, which contains high levels of mineral salts, tends to leave mineral deposits on plumbing fixtures and clog flush holes in toilets. Both groundwater and surface water pick up bacteria as well, but surface water picks up more. All these various impurities must be removed from water before we can start guzzling it.

Unnatural Cycles

Left to its own devices, the water cycle with its rivers, streams, and oceans would hum along just fine. We enter the picture and alter the cycle, sometimes to our detriment, by building dams, levees, and wells. Some observers have claimed that the disastrous floods in North Carolina this past year were due in part to an accelerated draining of wet lands, which normally would soak up excessive rainwater. Elsewhere, water tables have been lowered and altered irreversibly. Water cycles just don't get it when we try to maintain golf courses in the middle of the desert.

Pass the Salt—Not!

Another source of potable water is desalinization, or the conversion of salt water and *brackish* water to fresh water through the removal of its salt. The more accurate term

What's That Thingamajig?

Brackish comes from the old Dutch word *brak,* for "salty." It means slightly or moderately salty water, as opposed to sea water, which is heavily salty. Water flowing into a lake may pass through a salt source, sometimes a basin and sometimes an area associated with salt mines.

Plumbing Perils

Ground water from a well should be tested regularly for impurities. This water can become contaminated by polluted runoff, especially in agricultural areas containing fertilizers and pesticides. Regular testing also will detect harmful bacteria.

is actually saline water or any water with a significant concentration of dissolved salts as measured in parts per million (ppm). Desalinization is an expensive way to go, though. Several different processes are used, depending on the facility. Salt water is frozen, steamed, evaporated, shot with an electric current, or condensed, depending on the technology. In areas of water scarcity, desalination might be a competitive source for potable water, but don't look for it anytime soon off the regularly rained-upon Northwest coast.

The Middle East is a big user of desalinization plants, so much so that there were Y2K concerns about the plants computers. Oil is great for income, but you won't get far in the desert without water.

Giving Water the Treatment

Human beings like clean water. We don't just want the big stuff removed, such as fish, seaweed, and old tires, but we want to get rid of all the impurities, whether they're visible or invisible. This is where water treatment facilities and their counterparts, waste treatment facilities, come in. Early plumbing systems emphasized bringing water into populated areas. Waste removal, on the other hand, was pretty much approached as out of sight, out of mind, which didn't work too well for anyone downstream.

Water treatment plants vary in their complexity, depending on the condition of the water and local regulations. Treatment can range from chlorination (the addition of chlorine to kill germs) to screening, filtration, the use of activated carbon to remove odors and objectionable tastes, and even irradiation. It's really quite a science that only a sanitary engineer could love.

You and Your Water Company

Your local water utility is responsible for distributing fresh, clean water for residential and commercial use. In rural areas, unless a homeowner or farmer is supplied by a well, the utility also provides water for irrigation. According to Microsoft Encarta, the average daily water consumption in the United States ranges from 100 to 250 gallons per person per day. These figures include flushing the toilet, brushing your teeth, washing your car, and swigging espresso.

Water quality standards pretty much follow those set out by the United States Public Health Service. They determine the levels of bacteria and chemicals allowable in potable water. A utility might add certain chemicals to a water supply depending on the condition of the water and its effects on pipes. The most famous additive is fluoride, which has been used in local water systems since the 1950s.

Plumbing the Depths

Fluoridation, the adding of fluoride to the water supply, has been viewed as everything from a communist plot to a remarkable public health measure. Fluoride is a naturally occurring element that helps prevent tooth decay, according to its supporters. Small amounts of fluoride are naturally present in water supplies. The U.S. Public Health Service has recommended that additional fluoride be added to water supplies in the range of 0.7 to 1.2 parts per million to reduce tooth decay. Ironically, some home water filters can reduce fluoride levels in tap water, and bottled waters can be deficient in fluoride. Dental Luddites still argue against fluoridation on health grounds, and a Web search on the subject will call up a lively debate.

You might not want any of these in your water, but affordable, absolute purity is tough to come by and is unnecessary. Despite naysayers, our water supply is safe and well-monitored.

One City's System

Every water utility system follows the same functions: store and distribute potable water to its customers. The size and complexity of the customer base determines the means of distribution. The city of Davis, California, for example, describes its system as follows, according to its Web site:

> The City's Public Works Department maintains the water supply and distribution system for the City. The system consists of 21 water supply wells, 1 elevated water storage tank with a 200,000 gallon capacity, and more than 145 miles of water distribution piping ranging in size from 6 inches through 14 inches. The supply system produces an average of 11 million gallons per day. The production capacity is considered adequate to supply the current demand with sufficient reserve to meet peak demand and fire demand requirements. The City is proceeding with plans to construct a four million-gallon water storage tank by

1998, followed by a second tank projected for 2000. Twenty of the wells are operated by electric motors, and one well is operated by a natural gas engine. Two portable generators are currently available for standby power.

Water utilities employ reservoirs, pumping stations, storage tanks, and water towers to distribute water. The utility obtains water from an underground source or a surface body of water, cleans it, and pumps it to storage tanks. The water then is distributed through a system of pipes to residential and commercial users. Low-tech gravity plays a role in this distribution, too.

Plumbing Perils

Bottled water isn't necessarily "better" for you than tap water. It depends on the source of the water, its testing, and treatment, if any. The 1996 Safe Drinking Water Act requires that bottled water be tested to meet the same regulations and standards as tap water, which has been subject to rigorous standards for years.

Watering Heights

Water towers are elevated, neighborhood storage tanks. You see them in small towns all the time, where they regularly become targets for graduating high school seniors' graffiti skills. The height of a water tower provides pressure to push the water out and on its way. Locating a tower on a tall hill or some other high ground assures sufficient pressure to serve water customers.

A water tower's tank can hold a million gallons of water or more, typically enough for a one-day water supply to the customers served by that tower. Treated water is supplied to a water tower by the utility's pumps. High-rise buildings have their own pumps and water storage tanks to assure tenants constant water pressure.

From Underground to Overhead

A reservoir is a lake, often man-made, for storing and collecting water for our use. Reservoirs can hold water for agricultural use or for treatment for human consumption. Most reservoirs, at least in the United States, are also used for recreational purposes such as fishing and boating. Some jurisdictions prohibit any contact recreation such as swimming, and others prohibit any human activity to avoid possible contamination. In the eastern states, reservoirs might be less than 100 acres in size, while in the western states they can be measured in square miles. Many are managed or owned by the U.S. Army Corps of Engineers, while others fall under the auspices of other governing authorities.

The Inside Story

Getting water to your home is one part of the process. The next step is to distribute it where you want it in a usable and safe fashion. Plumbing is more than connecting a pile of pipes and opening the taps. Pipes have to be sized for sufficient water pressure and flow, valves must be installed to shut off water, and, equally important, a drain-waste-vent (DWV) system must be in place.

If you could see behind your walls and under your floors, you would find the following plumbing components:

➤ Cold water branch lines

➤ Hot water branch lines

➤ DWV pipes

Both cold and hot water pipes are under pressure and must be connected to exacting standards to prevent leaks. DWV pipes also must be connected properly, but because they are not under pressure, leakage is a little less problematic. It all starts with your main water supply pipe.

Water Enters Here

Water from your local utility is carried to your street through a large pipe called a water main. This pipe is sized to carry enough water at sufficient pressure to serve all the residential and commercial lines that will branch off it into individual homes and businesses. These water supply pipes are smaller than the main, of course, because individual customers demand far less water than a main can supply (teenage showering and bathing habits notwithstanding).

New homes typically have at least a 1-inch diameter water supply pipe, but this isn't always the case with older homes—they might have smaller supply pipes that don't always keep up with modern lifestyles and their water demands. If you add a bathroom to an old house, you'll feel a drop in water pressure as other fixtures in the house are used.

Your main water supply line, sometimes called a service line, carries water from the main through a water meter, which measures your usage. The meter might be located in your house, but often it is near the street near the *curb stop*, which shuts off the water entering your house. The supply pipe is buried beneath the frost line, so it shouldn't freeze in the winter, although it can freeze where it runs through any exposed sections of an unheated crawlspace or cellar.

z z z

Pipe Dreams

The first job in upgrading an old plumbing system is to replace the main water supply line if it's undersized. This is not a terribly costly job, and it will make a big difference in your water pressure.

What's That Thingamajig?

A **curb stop** shuts off water before it enters a house and is owned or controlled by your water utility. A **main shutoff valve** is located inside your house, under your control, and shuts off water before it goes to fixtures and appliances.

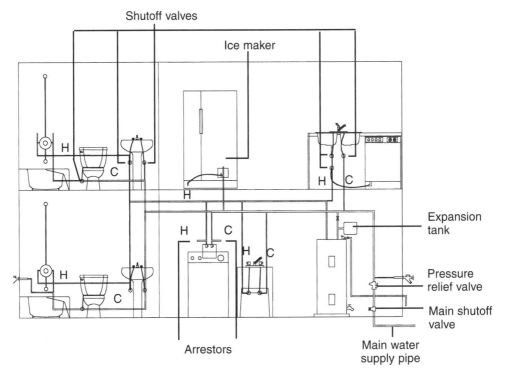

The water supply system.

Branching Out

When a water supply pipe enters your house, it passes through a *main shutoff valve*, which controls the flow of water before it goes to any fixtures or water appliances. The supply pipe then feeds both the hot water heater and all the cold water branch lines. The cold water lines feed each fixture and water appliance, including these:

➤ Sinks

➤ Tubs and showers

➤ Toilets

➤ Bidets

➤ Washing machines

➤ Dishwashers

➤ Ice makers

➤ Hot water heating systems

Hot water branch lines run parallel with the cold water lines to every fixture and appliance demanding hot water. Branch lines are smaller than your main supply pipe, either one-half inch or three-quarter inch in diameter. Like all pipes, they are sized to adequately meet the demand of the end fixture and its user. A fire hose, for instance, requires much more water than your kitchen sink and therefore has a much larger pipe supplying the hydrant with water (and a good thing, too, if it's your house that's burning).

What Comes In Must Go Out

All the water that comes to your fixtures and appliances has to go somewhere. The drain-waste-vent (DWV) system is arguably more important than the water supply side of your plumbing system. You can always bring water in by the bucket if you have to, but getting rid of unhealthy waste in a safe and effective manner is more difficult.

In older homes, DWV lines are made from cast iron, copper, lead, or galvanized steel. Plastic pipe rules in new homes and sometimes is used for supplying water as well. Each plumbing fixture and appliance is connected to the DWV. Their drain pipes (and waste lines for toilets) then connect to the soil pipe or main stack that empties into the sewer line. The main stack, along with other main vents, also allows sewer gas and odors to pass outside your house, a critical role for your health and well-being. In addition, these main vent pipes, which extend a minimum of six inches through the roof of your house, allow air to enter into them, and maintain an equalized pressure inside the drain pipes. Without this pressure, water inside the traps (see below) would be sucked out by a vacuum action, allowing sewer gas, which is nasty stuff, to enter your house through your drain openings in sinks, tubs, and toilets.

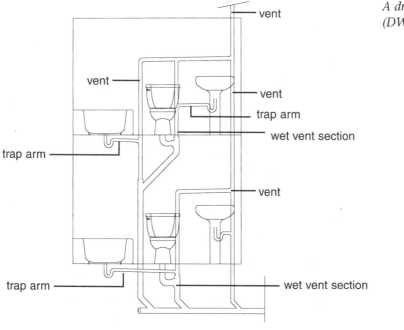

A drain-waste-vent (DWV) system.

Feeling Trapped?

Your DWV pipes are connected to your sewer line, a great source of noxious fumes collectively referred to as sewer gas. Fumes will be fumes, and because your sewer line isn't constantly filled with a flow water and waste, they will migrate into your home through the same pipes that carried your waste away. To prevent this, the DWV system has a series of ingenious yet simple drain traps built into it that prevent gas and fumes from entering your home.

A trap is a curved section of pipe, located under each fixture and built internally inside toilets. Water held in the trap seals off the sewer gas. Every time you run water or flush a toilet, this water is replaced so that the trap is always filled. A trap stays filled because the system is vented.

Each fixture is connected to a drain line that is in turn connected to a vent line. The vent lines connect to the main soil pipe or other main vents that pass through the roof of your house. Where the vents exit through the roof, they must be at least three feet above or ten feet to the side of any operable window or door. Each individual vent serves two purposes:

1. It vents any gases or odors that are present in the system to the outside.

2. It equalizes the water and air pressure in your waste lines.

Left to its own devices, water running down drains would siphon out any water in the traps. No water in the trap means no seal against sewer gas. The development of a working vent system was a *huge* advance for residential plumbing and didn't occur until the late 1800s. Chapter 2, "Potable Water, Potent History," covers the history of plumbing and some of its less savory moments.

All our waste water has to be removed, cleaned, and released by a sewage treatment facility. Before water can be pumped into our homes, it must be treated as well. As demand for water goes up, especially in areas with a more weather-dependent supply of water, conservation and wise water usage become more important.

Drain Dangers

Drains, including toilets, washing machines, and utility sinks, are great for carrying away unwanted wastes. The problem comes when we use them for waste that is harmful to the sewage treatment process. It's a good idea to keep in mind that you or someone else might be running into this same water again and again. Do you really want to pour paint thinner down the drain, too?

Most water utilities send out informational brochures from time to time listing the household products they *don't* want entering the waste stream. These products include the following:

➤ Any petroleum-based product, especially motor oil, paint, and paint solvents

➤ Insecticides and pesticides

➤ Excessive kitchen grease and oil

➤ Personal sanitary products

If you're unsure about a specific item, read the label or call your water utility.

Every Drop Adds Up

Water conservation isn't limited to arid regions of the country. If you've ever seen the casino fountains and the golf courses in Las Vegas, you'd wonder if they had ever even heard of water conservation, but that's another matter when illusion must trump reality for visitors. Many cities across the country have introduced water-saving devices such as low-flow shower heads to their customers and have implemented water restrictions during droughts. Even Seattle, land of the seemingly endless rain, has issued water restrictions when the snow pack and water levels have been lower than usual.

Systemwide conservation efforts can increase the reliability of water service by decreasing water demand, especially during extended periods of hot weather. From a utility's standpoint, it's a lot cheaper to promote household conservation and leak detection programs, and assist commercial users in decreasing their water requirements than building new treatment plants and infrastructure.

Household water conservation won't require a noticeable change in your lifestyle—it just means paying some attention to your normal water usage. Some conservation measures include these:

➤ Run only full dishwasher and washing machine loads.

➤ Install a low-flow shower head.

➤ Take short showers.

➤ Repair any leaks or drips in your plumbing.

➤ Don't leave unattended water running in sinks.

➤ Wash your car, or anything else outdoors, with a nozzle on the end of the hose instead of letting it run continuously.

➤ Consider installing drip irrigation in your garden.

➤ Keep a pitcher of cold drinking water in your refrigerator instead of running tap water until it's cool.

Pipe Dreams

A great way to control your water usage in the shower is to install a shower head with a manually controlled water flow. These have a sliding pin to cut the flow. You can decrease the water to almost a mist level, enjoy a longer shower, and still cut down the amount of water needed to shower.

Plumbing the Depths

Approximately 80 percent of Earth's surface is covered with water, but around 97 percent of it is saltwater contained in oceans; another 2 percent is frozen and found in glaciers. The old saying that they're not making any more land is really more appropriate for water, whose worldwide volume has remained essentially constant throughout time. The water you drink today is the same that was available to the dinosaurs. Freshwater lakes and rivers make up only slightly more of the water supply than do saline lakes. In the United States, we really like our water and use about 25 trillion gallons of it every year.

By the way, showering with a friend doesn't really save any water. It can be easily argued that it uses *more* water than showering individually, but you definitely will become friendlier. These other suggestions are all pretty easy to carry out and cumulatively will save you quite a bit of water over a year's time.

The Least You Need to Know

➤ Water might be everywhere, but it has to be cleaned, stored, and delivered before we can make much use of it.

➤ Reservoirs, water towers, and pumping stations are all critical parts of your local water utility.

➤ Pipes are sized, from your city main to your hot water branch pipes, to deliver water under steady, even pressure.

➤ Your drain-waste-vent system is probably more important than your water supply system, but you need both for a complete plumbing system.

Chapter 2

Potable Water, Potent History

In This Chapter

➤ Romans, our first big-time plumbers

➤ The Dark Ages kick in

➤ From outhouses to water closets

➤ America's attachment to bathrooms

➤ Sanitation rules!

Oddly enough, ancient cultures were *way* ahead of many present-day countries when it came to plumbing and sewers. The Romans, as we shall see, were big waterphiles and took their technology with them when they hit the road. Bath, England, was an appropriately named Roman outpost (in 50 A.D.) known for its baths, which were early versions of hot tubs. When the Romans fled Britain some years later, they took their technology with them. If you've ever seen Monty Python's *The Pursuit of the Holy Grail*, you'll know just how grubby the English became in the ensuing years.

Human beings have always needed clean drinking water for survival. As the great civilizations developed, water also was seen as an indulgence and comfort in the forms of baths, both public and private. Rudimentary waste removal systems were built in addition to heated structures for bathing. Cheap labor probably helped considerably.

Americans are probably the most bathroom-infatuated people in the world, and we've got the bathrooms to prove it. We're not the first, though, for history doesn't lack for efforts to bring water into homes and remove wastes. Plumbing systems and fixtures are more elaborate than ever, but the past can lend an interesting perspective and show just how far we've come in both areas.

When in Rome ...

The Romans were the first of the ancients to get serious plumbing rolling, but they weren't the first to install some kind of water management system. According to *Plumbing and Mechanical Magazine*, wealthy Babylonians had separate bathrooms in their homes with floors sloping toward a center drain. Extensive irrigation systems were created to control the flooding of the Tigris and Euphrates Rivers. And those famous hanging gardens had to be watered somehow, and they were with an elaborate watering and drainage system. The Egyptians also built elaborate bathrooms inside the pyramids so that their mummified occupants could relieve themselves on their way to the afterlife.

According to some researchers, the world's first "water closet," or toilet, was built in the Minoan Palace of Knossos on the Isle of Crete almost 4,000 years ago. Palace engineers constructed a drainage system with lavatories, sinks, and manholes for access to a masonry sewer. Terra cotta drainage pipes were flushed clean by rain water collected from rooftops and cisterns.

It wasn't an accident that royalty, the wealthy, and ancient priests were the first (and often the only) kids on their respective blocks with some form of a bathroom. Commoners just had to make do on their own, which is never the best policy for public health.

What's That Thingamajig?

A **water closet** is an old term for what we would now consider a small bathroom with a toilet and often a sink. It could also refer to the toilet itself. The closest term we now have would be a half-bathroom. An earlier generation or two would have used the term "powder room."

Plumbing the Depths

Manfred Klauda, of Munich, Germany, established what he believed to be the world's first chamber pot museum in the 1980s. Chamber pots, which go back centuries, were used both publicly and privately, it seems. Louis XIV apparently discussed government affairs while sitting on his pot. Klauda's collection of 5,000 chamber pots included his-and-hers earthenware pots for newlyweds and a nineteenth-century musical potty for children from Germany.

Bath Time

The Romans pretty much invented a systemized, large-scale approach to plumbing. Although the earliest pipes were made from wood or earthenware, lead eventually replaced them. The Romans really liked lead and used it for pipes and drainage systems. In fact, the word *plumbing* comes to us courtesy of the Latin word *plumbum,* for "lead." Roman lead workers were considered highly skilled and took care of both water supply and waste disposal systems.

The great Roman baths obtained their water via aqueduct from sources miles away and had both hot and cold water. The Spartans considered hot water to be unmanly, but the Romans were no fools—they apparently didn't see any virtue in meaningless suffering. One public structure, the Baths of Diocletian, seated more than 3,000 people and had ongoing streams of hot water by building controlled fires under the stone floors of the baths. A bath complex might feature several temperatures of water, including these:

➤ Frigidarium, or cold

➤ Tepidarium, or moderate

➤ Caldarium, or the hottest water

True marketers, the Romans apparently provided bath water for all tastes and temperaments.

The Party's Over

As civilizing as the Roman's plumbing system was, it didn't hold sway with the barbarian invaders who, one can presume, answered their personal hygiene needs whenever nature called. One can further presume that regular bathing wasn't on their daily "To Do" list. With the barbarians huffing and puffing across the continent, sanitation reverted back to the basics. Roman influence waned as the arena of ideas shrank. Early Christians, for example, considered cleanliness to be a sinful display of wealth, according to some historians. The barbarians, coupled with the destruction of Roman water systems, brought on a thousand years of squalor, sickness, and hardship.

As the Dark Ages set in, hygiene took a nose dive. As cities grew and population densities increased, sanitation and waste disposal became a growing problem. Without any form of a water closet or toilet, let alone a sewer system, chamber pots simply were emptied into the streets. London tenements were a horror, and the Thames River was essentially an open sewer as well as the city's chief source of drinking water. It got so bad that Parliament reportedly suspended its session for a few hot days as late as 1859 due to the stench. Attempts to mask the odor by saturating window blinds with disinfectants such as lime chloride were woefully unsuccessful. Waterborne diseases including typhus, typhoid, dysentery, and cholera spread. A few

What's That Thingamajig?

The **bubonic plague,** known as the Black Death, is transmitted by the bite of certain parasitic insects whose hosts are rodents. The rat flea Xenopsylla cheopis was the main infecting agent during the Black Death. When infected, a victim's lymph glands swelled; these inflamed lymph nodes are known as buboes. The 1666 Great Fire of London destroyed much of its source of the plague.

centuries earlier, such conditions were an open invitation to the *bubonic plague* to come and party hearty.

Reliving History

After some centuries of dirty living, Europe slowly took hold of its senses. Bath, England, was restored in the sixteenth century. Sir John Harrington invented his "Ajax" water closet in 1596 for Queen Elizabeth I. Unfortunately, it was subject to great ridicule, so he never expanded on or improved it, or even built another water closet. Some seventeenth-century castles had indoor privies, but they emptied into the moats (maybe this was to add to their defensive value).

Hit-and-miss attempts at inventing workable water closets in Europe, mostly in eighteenth- and nineteenth-century England, included the first float and valve flushing system invented by Joseph Brahma in 1778. This principle is still used in contemporary toilets (Please see Chapter 11, "Say Hello to Your Toilet.") The need for improved sanitation eventually brought the force of law into the area of plumbing.

Plumbing the Depths

According to Dr. Andy Gibbons, historian of the International Thomas Crapper Society (yes, there is such an organization), and researcher Ken Grabowski, "The Silent Valveless Water Waste Preventer" (No. 814), an invention that allowed for effective toilet flushing, was patented in Great Britain in 1898 by Albert Giblin. Thomas Crapper (1836–1910) had a successful career in the plumbing business but did not invent the flush toilet most often attributed to him. Gibbons and Grabowski suggest that Crapper bought the rights from Giblin for this greatly improved toilet and proceeded to market it through his plumbing business. Crapper himself held nine patents for different plumbing devices.

Parliament Cleans Up Its Act

In 1848, the Public Health Act was passed in England. With the force of the government behind it, this act mandated sanitary accommodations for every house, coupled with the beginning of a modern sewer system. This act became a world model that would be followed by similar legislation in the United States.

America and Its Plumbing

Early American plumbing systems showed, among other things, that we could do just about anything with all the free wood available in the form of seemingly endless forests. Logs were selected and bored out for use as pipes. Tradesmen called borers, appropriately enough, used a 5-foot long auger for the job. Individual logs then were joined at the seams with tar or pitch used as a form of caulking. Another approach was to split the logs, hollow them out, and then join the two halves with iron hoops or melted lead. Iron and lead pipes replaced their wooden predecessors in the nineteenth century.

It was unusual for an American colonist to have any kind of a private bathroom in the seventeenth century. Just having a privy or an outhouse was considered somewhat of a status symbol. If you think that an outside facility was inconvenient, especially in the winter months, you're absolutely right. Indoor plumbing of sorts was on its way, assisted in part by the invention of central heating in the 1800s.

Plumbing Perils

Oddly enough, Washington, the Evergreen State, still installed some wooden piping into the 1920s. If you think that wood and water don't mix, you're right. Ironically, the same company that installed these wooden pipes not too far from Seattle was hired to install modern replacements a few years ago.

Pay Attention to Your Infrastructure

Early colonists realized that they needed some reliable water supply in the event of fire. Boston built the country's first waterworks in 1652. The irony of depending on wooden water pipes was probably not lost on any of the more astute Pilgrim types of the time. Firefighters punched a hole in a wooden pipe wherever they needed water and later plugged the hole with a cone-shaped stopper on the end of a long pole. You guessed it—these were the original fire plugs (canine reaction to them is presumably unrecorded). In 1804, Philadelphia became the first city in the world to use iron pipe for all of water mains.

City water systems needed pumps, and early pumping systems—including water wheels and windmills—weren't going to cut it as urban populations grew. Coal-fired, steam-driven pumps eventually provided the millions of gallons of water necessary to supply increasingly industrialized cities and their populations.

With the water supply becoming more secure, the next step was getting rid of the wastes. Some historic views of the early settlers might see them as daring pioneers, ready to throw aside the social shackles of the Europe they left behind, but some traditions held firm, including tossing waste and garbage into the streets. Scavengers, with careers in the waste disposal business, were known to simply dump the contents of privy stations or outhouses in the streets. New York City had to pass an ordinance against such dumping as early as 1700.

It wasn't until the nineteenth century that modern sewer systems were developed in the United States. An engineer by the name of Julius W. Adams was commissioned to develop the sewers for the then-city of Brooklyn. His was a pioneering effort, and his guidelines and calculations, which he published for others to use, became the basis for modern sanitary engineering.

Privies Are Not Always Private

Well, people had to go somewhere. Outdoor privies, another term for outhouses, ranged from thrown-together wooden shacks to a brick structure with five holes built for the chief magistrate of the colonial court. The center seat was raised higher than the other surrounding four, as was befitting the judge's local status. Most Americans had more humble arrangements. According to The Outhouses of America (www.jldr.com/faqs.html), the typical outhouse was a simple wood structure that was moved to a new location when the hole was about full. The hole was covered with dirt while worms and bugs did their work on the hole's contents. Inside many outhouses there were two different-sized holes, one for adults and one for kids (who quickly figured out not to use the larger of the two). The familiar crescent moon on the outhouse door is believed by some to be an ancient symbol identifying an outhouse for women, but eventually became a unisex symbol that doubled as a source of light.

Plumbing Perils

Early outhouses, privies, and cesspools didn't entirely solve the problem of waste disposal. Well water, which many early Americans depended on, could become contaminated if these facilities were poorly built, ill-maintained, or badly located. It was a start, but municipal sewers wouldn't be on the horizon for decades.

Thomas Jefferson built an indoor privy at Monticello. His servants hauled out the chamber pots from their earthen closet below the privy's hole using a system of pulleys. Outhouses solved one problem, the need for some form of a toilet, but caused another: They could foul underground water sources.

Big Beginnings

In 1829, architect Isaiah Rogers designed Boston's Tremont Hotel, the first of its kind to have indoor plumbing and the prototype for first-class hotels. Eight water closets on the ground floor were fed by a large metal water tank on the rooftop, which was filled with pumped water via a steam pump. The water was drawn down by gravity to fill tubs and flush away wastes. In 1834, Rogers outdid the Tremont with his design for New York City's Astor House, implementing water closets and bathrooms to serve 300 guests.

The White House was technologically behind the times when it came to plumbing. This isn't surprising, given that up to and until the end of the George Bush administration, it still used switchboard operators to handle the phones. Say what you will about Bill Clinton—at least he upgraded the phone system at the White House.

During its original construction, the White House had no bathrooms; servants had to haul in water from a spring five blocks away. The home's first water pump was installed in 1833, when iron pipe was laid from a bubbling spring to a pumphouse. The first major overhaul of the plumbing took place in 1902, but the job wasn't done all that well. Hit-and-miss repairs and additions over the years compelled Harry Truman, who discovered that his bathtub was sinking into the floor, to authorize a multimillion-dollar reconstruction project.

Pipe Dreams

Early in the nineteenth century, the idea was born to vent the main stack through the roof. Although no one then knew how to size the stack properly and often undersized it, the science of venting became more accurate toward the end of the century.

Bringing Outhouses Indoors

Indoor water closets were certainly more user-friendly than outhouses, particularly in the winter. Early models were wood-and-metal affairs that remained a source of contamination because they could not be properly sanitized. The wood especially was a problem because it would rot away from contact with water and waste. The introduction of one-piece porcelain fixtures in the late-nineteenth century did much to solve the sanitation problem. Fixtures evolved from conical-shaped hoppers, pan closets, and wash-down closets with a water flush from an attic storage tank, to the plunger closet and trapless closet, both of which were considered to be unsanitary. William Smith produced a jet siphon closet in 1876, a design that eventually was improved upon by others.

The British produced superior water closets until the end of the nineteenth century, when American manufacturers caught up with new and improved porcelain fixtures. Once the fixture was exposed and no longer surrounded by a wooden box, manufacturers began decorating the outside and the exterior surface with pedestal patterns as well as the names of their companies. As demand increased, the range of fixtures increased, as did color choices and designs. Early water closets and plumbing efforts still were a matter of hit-and-miss guess work until plumbing standards were established in the latter part of the nineteenth century.

Bathing Was a Lot of Work!

Early American bathing often consisted of jumping into a nearby cold stream or lake, getting really cold, and drying off really fast. People became pretty rank during the winter months. Eventually, the idea of "taking in the waters" for their therapeutic value, as opposed to strictly cleaning, took root as spas developed around natural mineral water springs. At the same time, many considered bathing to be a health hazard, and some cities considered ordinances against it.

An added problem was the amount of work it took just to prepare a bath. Water had to be pumped into a pail by hand and then heated and poured into the tub, although some tubs had heating units attached to them. As the convenience of bathing improved, so did its frequency. Some tubs, for instance, were fitted with two pipes running from an attic water tank. One pipe supplied cold water, and the other coiled around a chimney and supplied hot water.

Early tubs were often lead-, copper-, or tin-lined wooden affairs. Cast iron eventually replaced these metals and was itself covered with a porcelain finish in later years for a more sanitary, easy-to-clean surface.

Pipe Dreams

Americans loved their plumbing gadgets from the very beginning. Shower heads date back to at least the 1840s, and all kinds of shower/bath combinations were created. George Hinman, of Portageville, New York, invented a seat for the water closet that flipped up and away when the user was finished. His invention might not have endeared him to women, however.

Does Plumbing Lead to Morality?

In academia, all subjects are fair game for research papers, dissertations, or books, and plumbing is no exception. Maureen Ogle, assistant professor of history at the University of Southern Alabama, took it upon herself to study the phenomenon of nineteenth-century American residential plumbing. Her book, *All the Modern Conveniences: American Household Plumbing, 1840–1890*, most likely never will make the list of best-sellers in *The New York Times*, but it's an interesting, if not dense, study of plumbing in transition and the reasons it was installed in the first place.

Ogle contends that early public water systems were not installed for household use, but primarily for protection from fire. City officials were even dismayed at the amount of water going toward household use as people brought it into their homes. Early residential plumbing systems depended on their own water supplies in the form of wells, cisterns, storage tanks, and pumps. Some made arrangements to pump water from a neighbor's land or simply ran pipes to a nearby stream or lake. Public water and household water were viewed as separate affairs.

Bacteriology and knowledge of germs as a cause of disease was unknown until the latter half of the nineteenth century. The incentive for indoor plumbing, according to the author, wasn't sanitation, but more an interest in self-improvement. Self-help advocates advocated the importance of convenience to a quality domestic life. Convenience was equated with good health and moral improvement; the elimination of drudgery, such as hauling in water by the bucketful, would ensure the well-being and good health of women, who in turn would devote all their efforts to nurturing the family. Somehow, guys conveniently got left out of the equation, but what else is new?

Because indoor plumbing was a convenience, it therefore was morally desirable.

Plumbing the Depths

Professor Ogle reports that one New York plumber's fees for plumbing work in "an average house" in 1840 were $600, and closer to $2,000 for "a very fine house." A sum of $2,500 got a homeowner 5 bathtubs, 1 hip bath, 5 water closets, 14 basins, a copper boiler, wash trays, and a copper butler's sink. One 1833 price book and estimator lists a plumber's fees at $2 a day. By 1855, the fees had increased to $3.50 a day. Given the amount of inflation during the last century and a half, these prices weren't at all cheap, a common lament still made about plumbing costs.

Self-Sufficiency Has Its Limits

Early plumbing systems relied on water storage in the form of *cisterns* located either in the attic or over individual fixtures. Water would flow through pipes to sinks, tubs, and water closets. The cisterns were filled by pumping water into them or, in the case of pumpless homeowners, by carrying it upstairs by the bucket, an arduous and time-consuming task that hardly qualifies as convenient. Some architects disdained attic cisterns because of the strain their weight put on the house and because they tended to leak. Some questioned the wisdom of using lead linings as a storage system for water (others were lined with zinc and even slate). As city water gradually became available, cisterns were abandoned.

Waste elimination was still the province of the homeowner, which meant cesspools and privy vaults, often built and maintained with minimal government regulation.

Also, there were no established standards for the practice of plumbing or the manufacturing of fixtures. Consequently, fixtures leaked due to both their design and installation procedures. Tubs, which were often wood-framed and metal-lined, split and rotted, as did the floors underneath when the wood soaked up water. Rats also ate through lead pipes, and the pipes themselves froze in the winter despite attempts to insulate them with sawdust.

What's That Thingamajig?

A **cistern** is simply a reservoir or storage tank for water. Early plumbing systems depended on gravity to move water, so overhead cisterns were necessary to store water and give it sufficient height to move forcefully.

The Sanitarians Arrive

In 1870, the New York Metropolitan Board of Health issued its Metropolitan Health Law, a model piece of health legislation. The law encompassed the study of ground water drainage, sewage and waste disposal, water supply, and the characteristics of water closets. Plumbing could only improve as a result of this legislation. Plumbing health codes were written to encompass installations as well as the training, examination, and licensing of plumbers.

The incipient sanitarian movement of the 1870s shunned past plumbing practices as unhealthy and focused on disease and the need for plumbing standards. When residential plumbing was tied to sanitary sewers, the era of private, self-contained systems came to an end. Specifically, the sanitarians pushed for three improvements:

➤ Standardized installations and training of plumbers

➤ Research

➤ Improved ventilation and plumbing traps

By the 1890s, Americans had begun looking at plumbing from a health and sanitation standpoint. New municipal regulations regarding public water utilities were enacted. Pottery manufacturers were producing inexpensive, sanitary fixtures, replacing the wood- and metal-based fixtures of the previous decades. Manufacturers formed organizations to spur the standardization of fixtures and components, something we take for granted today.

Life Is Better Today

Nostalgia for the past rarely deals with the whole picture. Life might have been slower and neighbors might have known each other better, but people in the past also had to put up with cold houses, were fatally struck by illnesses we can now cure with a pill, and lived with terrible plumbing. Sanitary water and sewer systems are absolute hallmarks of civilization, both of which were developed in this country. Okay, maybe we're a little obsessed with bathrooms—where else do you find houses with more bathrooms than bedrooms?—but there are worse cultural extremes to contend with.

Plumbing the Depths

According to Roto-Rooter, Samuel Blanc finished inventing his first machine for sewer cleaning in 1933. He built it from a Maytag washing machine motor, roller skate wheels, and a $3/8$-inch cable to turn the special blades or knives used to cut tree roots inside sewer lines. Depression-era entrepreneurs paid $250 for each machine and set up their own Roto-Rooter businesses. The big advantage of this machine was its guaranteed capability to clean out roots without requiring that the line be dug up. Today, Roto-Rooter is synonymous with drain cleaning.

The Least You Need to Know

➤ The ancients had surprisingly sophisticated plumbing and sewer systems; unfortunately, most of this knowledge was lost as their civilizations declined.

➤ Until the late nineteenth century, plumbing was a hit-and-miss affair with both homeowners and municipalities.

➤ Early water closets solved the question of where to go, but they didn't resolve the final question of disposal.

➤ Sanitary water and sewer systems made huge strives in the eradication of disease, especially in large cities.

Chapter 3

Codes, Inspections, and Safety

In This Chapter

➤ Your local plumbing code

➤ The importance of an inspection

➤ Problems with uninspected work

➤ Safe work practices

Union plumbers go through a rigorous apprentice training program. They are trained in the basics with an understanding that plumbing, sanitation, and human health depend on accurate, safe practices of their craft. Your plumbing repairs won't require the same extensive knowledge that a trained plumber will have, but you must be just as attentive to the quality and code requirements of your work.

Plumbing isn't like roofing: If you fall in the bathroom, it's nothing like tumbling off the peak of your roof. Nevertheless, there are hazards associated with any kind of repair or remodeling work, and you need to be aware of them. In a world increasingly oriented toward working at a desk, there is a growing lack of familiarity with more physical labor and the accidents that can occur. This chapter acquaints you with common sense safety practices so that you'll be able to enjoy the fruits of your labors instead of nursing an injury.

This Isn't Computer Code

Your local plumbing code is a set of rules that you must follow for your work to be safe and comply with local regulations. Plumbing codes are a good thing: They help ensure that our water is clean and that we are protected from unsanitary conditions. The code affects the delivery of all fluids and gases through pipes for both residential and commercial applications, as well as the removal of waste and sewage.

Many, but not all, plumbing jobs require that you take out a permit from your local building department. A permit states that you may proceed with certain defined work that will then be subject to an inspection. The inspection checks and confirms that the work is done according to the code. Performing the work without a permit or inspection can bring you all kinds of grief, regardless of whether you do the work yourself or a plumber does it.

National Plumbing Codes

The final arbiter of your plumbing work, and the agency whose rules you must follow, is your local building department. These codes are based on one of several national plumbing codes. The specific model code used depends on where you live. This is unlike the National Electrical Code, for instance, which is the single standard across the country on which local codes are modeled. National codes are advisory, while the local codes are enforceable.

Some of these codes include the following:

➤ The BOCA National Plumbing Code (established by the Building Officials and Code Administrators International, Inc.), used in the Midwest and Northeast.

Plumbing Perils

A major bathroom remodel requires separate inspections for plumbing, electrical, and any structural work. Don't assume that the plumbing inspector will cover any area other than plumbing.

➤ The Standard Plumbing Code (established by the Southern Building Code Congress International, Inc., or SBCCI), used in the Southeast and Southwest.

➤ The National Plumbing Code (established by the National Association of Plumbing, Heating, Cooling Contractors), used in Maryland, New Jersey, and some cities.

➤ The Uniform Plumbing Code (established by the International Association of Plumbing and Mechanical Officials), used in the western United States.

➤ The International Plumbing Code, a new plumbing code established by a collaboration of the International Conference of Building Officials

(ICBO), BOCA, and SBCCI memberships. (This code is considered by some to be more flexible than the Uniform Plumbing Code.)

These codes serve as models for state and local plumbing codes.

Local Codes Rule

Codes might even vary within your state, so don't take anything for granted. One of the major code differences among municipalities is the acceptance of certain plumbing materials. One might accept a certain form of plastic pipe, but another might not. Among other things, plumbing codes cover the following:

➤ The regulation of materials

➤ Connections of pipes and of drain lines

➤ The drain-waste-vent system

➤ Plumbing fixtures

➤ Hangers and support for pipes and drain lines

➤ Hot water tanks

➤ Gas pipe installation

Be sure to check with your city or town before you undertake a plumbing project to make sure that everything will be up to code.

Plumbing the Depths

A 1999 report in *The Boulder Daily Camera* stated that the state legislature was considering a bill "that would allow local governments to select their own plumbing codes and approve new plumbing materials." Any local code would have to be based on a nationally recognized code. House bill 1145 also would allow local governments latitude in approving alternative plumbing materials instead of having them reviewed by a state plumbing board. Some plumbers and trade associations objected to the bill, questioning the reliability of alternative materials and the need to know all the different city codes. Plumbers were concerned because they are ultimately responsible for the materials and fixtures they install.

One Code's History

A mechanical code doesn't appear overnight. It takes years of data collection and collaboration among different industry groups, including unions, manufacturers, and state health organizations, to produce a usable code. A code isn't set in stone, either: New materials and techniques are constantly evaluated and added to subsequent editions of the code.

The International Association of Plumbing and Mechanical Officials (IAPMO), the group that developed the Uniform Plumbing Code, began in 1926 as the Plumbing Inspectors Association of Los Angeles. The organization gradually gained recognition as it developed standards for plumbing installations and inspections. Its first published codes, in 1932, were the Standard Plumbing Code, the Standard Gas Code, and the Standard Water Pipe Code, all of which later became the single Uniform Plumbing Code. The organization's efforts culminated with the publication of the 1997 Uniform Plumbing Code, which was, in the organization's own words, "an amalgamation of the most desirable aspects of the three most respected plumbing codes in existence":

➤ The 1994 Uniform Plumbing Code, published by IAPMO.

➤ The 1993 ANSI A40 Safety Requirements for Plumbing, published by the MCAA and NAPHCC Joint Task Force.

➤ The National Standard Plumbing Code, published by NAPHCC. The IAPMO said, "For the first time in history, a plumbing code has been created which is the result of a collaboration of industry-wide entities."

Developing this code was no small task, and the public health is better off for its existence. The bureaucratic tedium necessary to produce this document—thousands of hours of meetings and conferences, data gathering, and correspondence—is immeasurable. Meanwhile, some of these same organizations were developing other codes, such as the Uniform Mechanical Code to cover heating, ventilation, and air conditioning, and even the Uniform Swimming Pool, Spa, and Hot Tub Code (the IAMPO is headquartered in California, after all).

According to the IAMPO's statistics (www.iapmonet.org), the Uniform Plumbing Code has been adopted by 14 states as an alternate to their own individual codes and is used in a total of 34 states. A lot of work went into developing the guidelines behind every faucet installation or toilet replacement you do in your home.

Permits and Inspections

Generally speaking, you need a permit whenever you alter your pipes or drains by extending them or adding to the system. This includes changes such as these:

➤ Building a new bathroom

➤ Installing a dishwasher

➤ Adding a laundry room

➤ Relocating or adding hose bibs

➤ Installing a water heater

➤ Adding a gas fireplace or appliance

Repairs may or may not require a permit, depending on how much you're doing. Check with your local building department for details. The city of Canton, Ohio, for instance, states the following:

> Permits shall be required for the installation, repair, and replacement of all plumbing work done in the City, or for plumbing work connected to the City's municipal water supply and/or sanitary sewer system beyond the City's corporation line … (Chapter 1315 of the City of Canton Codified Ordinances) including, but NOT all inclusive, fixtures or traps, waste piping, water piping, gas outlets, water heaters, food grinder, disposal or dishwasher, water service, water main distributing systems, and inspections. All contractors performing plumbing work in the City of Canton and all journeymen and trainee/apprentices must also be licensed.

If a permit and inspection are required for your work, you'll have to do the following:

➤ Apply for the permit and pay a fee.

➤ Have the *rough-in* stage of your work, if any, inspected.

➤ Schedule a final inspection for the completed job.

Permit fees vary depending on the size of the job. A basic fee covers most small jobs, and it increases from there. A job that requires installing new pipes in the walls will need one inspection to pass this rough-in work and a second one after all the fixtures have been installed. This is the finish or trim stage of the job. Inspectors can't stop by on a

Pipe Dreams

When in doubt, choose a conservative approach to your plumbing, one that's sure to pass inspection. It might mean some extra work on your part, but it's worth it if you pass the inspection without any hassles. You don't want to be arguing code with a plumbing inspector.

What's That Thingamajig?

The **rough-in** phase of plumbing work is the installation of pipes, drains, and vents within the walls or floors. This work must be inspected before the wall and floors can be covered and finished. A second inspection looks at the fixtures after they are installed.

whim; they must be scheduled. During the summer, when there's a lot of remodeling activity, it might take a while to schedule an inspection. If you schedule prematurely and are not going to be ready at the appointed time, cancel the inspection. This is just basic courtesy and keeps you on your inspector's good side.

The Code Isn't Perfect

Sometimes plumbing manufacturers follow the example of software manufacturers by introducing what are essentially beta versions of their products: They've had some testing, but maybe not enough in the real world to truly predict their reliability. A case in point is plastic pipe.

Plastic pipe comes in various flavors, including these:

➤ Polyvinyl chloride (PVC)

➤ Chlorinated polyvinyl chloride (CPVC)

➤ Polybutylene (PB)

➤ Acrylonitrile butadiene styrene (ABS)

PVC is used just about everywhere for DWV systems. It's lightweight, easy to handle, and available from both plumbing suppliers and home improvement centers. Unlike old galvanized pipe, PVC will never corrode. CPVC is used for hot and cold water supply but might not be permitted by your local code.

Pipe Dreams

Plenty of Web sites cover ABS and PB pipe, describing the pipe and how to identify questionable production runs from different companies. Numbering systems are listed in addition information on lawsuits. If you're not sure about your pipes, these Web sites will give you the information you need.

PB pipe is flexible and used for water supply, at least when it isn't leaking. Complaints were so numerous and severe that a class action suit in Texas against the manufacturers of PB ended with a judgment amounting to hundred of millions of dollars in favor of the plaintiffs. Hundreds of thousands of homes have been piped with PB in part because of the cost savings over installing copper pipe. Manufacturers have claimed that there's nothing wrong with the product and that the leaks are an installation issue. Given the diverse contractors installing this pipe, however, it's unlikely that the plumbers are all to blame when it fails. Regardless of who's at fault, though, some cities have banned its use—and there is that pesky legal action, of course.

ABS isn't used to supply water, but to drain fixtures. Accusations of deterioration also have resulted in a huge class action suit, which doesn't claim that all ABS pipe is defective, just the pipe supplied by the named defendants over a specific period of time. Both ABS and PB were approved by plumbing codes, yet they haven't stood the test of time.

What an Inspector Expects

An inspector looks at a do-it-yourselfer's work with more scrutiny than a plumber's job, although he or she will look at both of them carefully. An inspector wants to see a neat job done to code. Be sure that your job site is clean and picked up: This is viewed as a sign of careful workmanship. Regardless of which phase of an inspection you're calling for, have all that work completed; an inspector can't sign off on anything that isn't done.

The inspector has the final word. If you have any questions about your proposed job before you start, bring them up with someone at your local building department. You also can procure a copy of the code from them.

Inspectors Aren't Perfect, Either

It might come as a surprise to some plumbing inspectors, but they make mistakes, too. In our own house, which was built in 1994, the inspector failed to notice that an exposed gas line didn't have enough supports securing it to the garage ceiling. We added supports after we moved in, but the inspector should have caught this. You should be sure that you accompany your inspector during the inspection in case you have any questions about your own plumber's work or even your own. You want to find out sooner rather than later if you have a problem installation.

Plumbing Perils

Plumbing advice from neighbors, hardware store clerks, or home improvement center handouts is just that: advice. It isn't necessarily fact. Your local building code is the only source you can depend on, so check anything you hear against it before you start your work.

Uninspected Work

Watch Out! I'm not going to pretend that all plumbing work is done with a permit and follow-up inspections. That's like saying that everyone's tax return doesn't fudge on a deduction now and again. Plumbers and homeowners alike do repairs and remodeling every day that legally should be inspected but are not. How big a problem can it be if the work is done properly and to code?

The problem isn't the good work, but the bad work or the work that's done incorrectly because of a misunderstanding of the code. You could have an ongoing leak inside a wall and not know it until a lot of damage has been done. Pipes buried outside in the summer could freeze in the winter because they weren't buried deep enough. The wrong pipe size might mean you never have adequate water pressure.

And then there are the fines. If it's discovered that you or your plumber are doing plumbing work without a permit, you'll not only have to pay for a permit, but you'll have to pay a penalty on top of that. You can plead ignorance, but a plumber cannot. As the homeowner, too, it will ultimately land in your lap.

Plumbing the Depths

Uninspected work doesn't go away when you sell your house. One physician I knew in Seattle sued the former owner of his house when he discovered that an addition to the house had not been done with permits and was not done to code. Inspectors didn't catch it when the house was purchased, but someone did when the doctor went to sell the house a few years later. It was a mess for everyone, including the real estate agents involved with the original sale.

If nothing else, consider what can happen if you have uninspected work done and a pipe bursts and leaks, sending water down from the second floor to your basement. Normally, this damage would be covered by your homeowner's insurance, but insurance companies don't have to pay for negligent work (check your policy). The lack of a permit and inspection can cost you thousands of dollars in repair costs: Running water can damage drywall, wood floors, carpet, wiring, and furniture.

When you sell your house, you must disclose any work that was done on it while you were the owner. You also must disclose whether any of this work was done without a permit. This can become a sticky negotiating point, especially in a buyer's market. If you do not disclose and the next owner incurs damages as a result of faulty work, you could end up paying for them.

Few of us are fond of government regulation, at least not until it suits our own needs. If you need a permit, get one. Think of it as self-protection without having to take a bunch of martial arts classes.

Safety Counts

More construction companies are hanging banners across their job sites touting their safety record and their emphasis on job safety. Years ago, when the industry was more rough and tumble, this wasn't always the case. Professionals now emphasize safety, and so should you.

Any remodeling project can produce an injury, whether it's from a wrench that accidentally slipped off a fitting or because you fell into a sewer line trench. It's not just your safety that you should be concerned about, either, but that of other workers, your family, and anyone else on the job site.

Possible hazards when working with plumbing include these:

➤ Hand and power tools

➤ Faulty extension cords

➤ Incorrect lifting of tubs and toilets

➤ Debris, metal shavings, and dust

➤ Lead pipes

A few common sense precautions should keep you and yours intact and away from the first-aid kit.

Plumbing Perils

Children and pets are particularly vulnerable during remodeling projects. Be sure that they're kept out of harm's way and that power tools are unplugged when not in use. Keep any holes in the floor or outside covered with plywood, and keep doors to torn-up bathrooms closed.

Don't Fool With Tools

In the United States, we pride ourselves on our adaptability and willingness to break the rules. Unfortunately, this includes using wrenches to hammer nails and using pliers when we should use wrenches. Use a tool for the purpose it was designed, and you'll get better results without banging up your hands in the process.

Sharp tools—saw blades, drill bits, and utility knives—work best when they're sharp. Dull blades and bits can slip and slice into fingers and hands. Check that the teeth in your pliers' jaws aren't worn smooth and that hammer heads hold tight to the shaft. Don't use screwdrivers as chisels—that's a great way to round off their tips and guarantee that they will slip off the head of a screw. Use a chisel when you need one, not your screwdriver.

Pipe Dreams

Wear work gloves when you're using hand tools, but take them off when you're using power tools. The gloves can get caught in spinning drill bits or moving saw blades. Also use gloves when installing pipe or soldering joints.

How Shocking

Electricity and water don't mix, but you need power tools if you're going to drill and cut through wood and plaster to install your plumbing. You could always relive the old days and use hand tools, but plumbers doing the work back then probably dreamed of power tools to make their labors easier. Using power tools carefully will keep you safe and extend the life of the tools. Careful power tool practices include these:

➤ Check the cord for tears or cuts in the insulation.

➤ Never remove the grounding pin from the plug when one is present.

➤ Inspect the case for cracks.

➤ Replace broken triggers.

➤ Never bear down excessively on a tool when in use—this can strain the motor.

➤ Use the power tool in dry, not wet, conditions.

➤ Lift a tool by its handle, not by its power cord.

➤ Don't drop the tool, especially from the second floor to the first.

➤ Put your tools away when you're finished with them.

➤ Wear eye protection.

Plumbing Perils

It's a good idea to install a ground-fault circuit interrupter (GFCI) in your bathroom and kitchen. This special receptacle cuts off power in the event of a shock before you can become badly injured. One GFCI adapter plugs directly into an existing receptacle without requiring any wiring.

These are the nagging parent rules, but that's because your nagging parents had to buy their own tools (and appliances and cars) and wanted them to last. Yours will last, too, with a little diligence on your part.

Extension Cords

We all use extension cords, and we often abuse them. Extension cords have a stated ampacity (the capability to carry a certain amount of current) depending on the size or gauge of their wires. A 12/2 extension cord will handle all your power tool needs with current to spare. A cord that's too small can get dangerously hot and become a fire hazard if your tool has a higher rating than the cord and demands more electricity than the cord can safely supply. You can't go wrong with a larger 12/2 cord.

Extension cords should be stretched out when in use and not kept coiled (they can conceivably overheat this way). Check yours for tears and cracks, and be sure that the grounding pin is intact.

What's That Thingamajig?

A 12/2 **extension cord** means that the cord has 12-gauge wire and includes one hot conductor and one neutral conductor, just like household wiring. Modern cords also have a grounding conductor. The higher the wire gauge (14/2, 16/2, and so on), the lower the amount of current that the cord can safely carry. Twelve-gauge wire can handle all small power tools without the threat of overheating.

Going Up

Toilets are awkward, and some old fixtures—especially tubs—are heavy. Lift with your legs, not with your back bent over. If you're not comfortable lifting a fixture alone, get a partner or helper to give you a hand. There are no heroics in injuring your back (unless you're trying to get out of doing more work).

These Aren't Dust Bunnies

Any airborne particle—sawdust, plaster dust, metal shavings, insulation—is an irritant. Insulation fibers are the worst, in my opinion. A variety of dust masks are available, from inexpensive paper masks for light dust all the way to full face respirators with HEPA filters for removing even the finest dust. For this type of work, a 3M 8710 disposable mask should be sufficient for most nuisance dust except asbestos. Wear eye protection as well. Protective safety goggles cost only a few dollars, and you should always keep a pair in your tool box.

The Evils of Lead

Some older homes might still have some lead pipes and closet bends. If you believed some observers, the use of lead in residential construction, particularly lead-based paint, is the most dire health hazard to which human beings have ever been exposed. It's pretty unlikely that you could ever ingest any of it by removing old pipes, but be sure to wash your hands thoroughly when you're finished working, especially before handling any food. You should call your building department and inquire if there are any special requirements for disposing of lead debris. The rules are different when a homeowner produces lead waste instead of a contractor.

Plumbing Perils

Your plumbing work might require that you remove a pipe that's wrapped in asbestos, some varieties of which are considered a health hazard and lung irritant if breathed. Some municipalities allow homeowners to do this, but others might not. Strict rules govern the removal and disposal of asbestos. Your building department can supply you with the appropriate guidelines.

The Least You Need to Know

➤ Your local building department can tell you whether you need a permit for your plumbing work.

➤ Skipping a permit and inspection when you need them can lead to problems down the road.

➤ You don't have the final word on your work—your inspector does.

➤ Safe tool practices will get the job done sooner and keep you from harm's way.

Chapter 4

If Your Walls Could Talk

In This Chapter

➤ Inspecting your house and plumbing

➤ Looking at the basics

➤ Plumbing questions you should ask

➤ Deciding your plumbing's fate

Whenever we walk into a house, we're struck by everything visual: the colors, decorations, furniture, and floor plan, among other things. We can't see inside the walls and know how well it's built. Did the builder use select, kiln-dried framing lumber, or lumber that was still a little green and more subject to shrinkage? The house might have been built to code, but was it built to exceed the code?

The older the house, the more problematic its construction. Telltale signs, such as cracks from settling, water stains on the ceilings, or amateur remodeling, will give you a clear idea of the physical condition of the house. A thorough inspection should always be done before you purchase a house, regardless of whether it's brand new or a turn-of-the-century Victorian. The same observational skills and questioning you bring to a prospective home purchase can also be brought to the home you're living in now.

Before you do any plumbing work (other than minor repairs), you should have a good understanding of your plumbing system. You should know what kind of pipe is used in your house, the size of your main water supply pipe (service line), and whether repairs and improvements done to the system were carried out properly. You can hire a professional inspector, or you can do it yourself. All it takes is a pen and a notepad to record your findings, a flashlight, and maybe a couple hours of your time, depending on the complexity and size of your system. When you're finished, you'll have the knowledge you need to do future remodeling and additions to your system.

Inspection Essentials

A house isn't exactly going to speak up and tell you that there's a leak in the chimney flashing or that the washing machine is on its last legs, but it can give you hints. The simple age of a house will tell you something about the building techniques and materials employed during its construction. A 1920s bungalow in mostly original condition, for instance, might have all or some of the following:

➤ Cedar siding

➤ Galvanized plumbing pipes

➤ Knob and tube wiring

➤ Single-pane wood windows

➤ Superior (by today's standards) grades of lumber

On the other hand, it might not have some of the following:

➤ Insulation

➤ Anchor bolts securing the framing to the foundation

➤ Copper plumbing

➤ A modern, grounded electrical system

Plumbing the Depths

As a rule, you don't have to bring your house up to the current building code unless your remodeling project requires code compliance in a certain area such as electrical or plumbing. Under some extreme conditions, local building authorities can demand emergency repairs. An undermined and sagging foundation, for instance, might be considered dangerous and render a house uninhabitable until it's repaired. Still, updating a house to meet modern codes is a good idea.

An old house, even with all the original mechanical systems, can be perfectly livable, but usually a house of a certain age needs some work. Plumbing normally makes the top 10 list of needed improvements. Even newer homes need to be scrutinized, especially in light of lawsuits over the installation and use of certain plumbing materials.

As a homeowner or prospective homeowner, you can do your own inspection or hire a professional.

Do-It-Yourself Inspections

If you're comfortable with your knowledge of construction, then by all means do your own home inspection. It should be a thorough process, including poking and probing in all corners of the attic and any crawlspaces. Besides looking for plumbing problems, you'll be checking for the following:

➤ Termite and other pest infestation

➤ Structural flaws

➤ Leaks in the roof

➤ Mechanical deficiencies in the electrical, heating, and air conditioning systems

➤ Drainage problems in the yard

➤ Overall condition of the house and yard

Ask yourself some questions:

➤ How familiar am I with the construction methods used in this house?

➤ How well can I evaluate the mechanical systems?

➤ Am I aware of recent recalls or consumer alert notices for different building materials, some of which might have been used in this house?

➤ Do I really want to crawl around under the house and climb around on the roof?

You might not know the answer to some of these questions until you do your inspection. If you find that you're in over your head and need a more knowledgeable opinion, consider hiring a professional inspector.

Inspector Homes, At Your Service

There are inspectors, and then there are inspectors. When we purchased our own house, even though it was new, we had an outside inspector come in for an objective opinion. After spending a whopping two hours for a fee I don't care to divulge, he came up with three items he thought needed to be corrected. He dutifully checked these off on his printed forms and was wrong about two of them.

That said, a knowledgeable inspector can save you a lot of headaches by pointing out problems before your purchase. Get some recommendations from friends who have purchased homes recently (they'll know who's competent and who isn't). Be sure that your inspector has been certified by the American Society of Home Inspectors (ASHI), an organization whose members participate in continuing education programs.

Plumbing Perils

Plumbers and other contractors normally give free estimates, but don't abuse this by trying to use an estimate as a cheap way to get an inspection of your system. Play fair—plumbers have to make a living, too.

What's That Thingamajig?

Coliform, according to The American Heritage Dictionary of Science, refers to the "bacilli commonly found in the intestines of humans and other vertebrates, especially the bacillus Escherichia coli." Yes, it's just as nasty as it sounds. Well water can get contaminated by livestock wastes from a neighbor's property. Rural well water should always be tested.

If you're inspecting only your plumbing, consider hiring a plumber to walk through for an hour or so. Your friends or neighbors should be able to give you the name of a plumber they can recommend to give you an objective opinion.

Regardless of who inspects the plumbing, that individual should examine the following:

➤ The source and condition of the water

➤ The water pressure and volume of water

➤ The age and type of pipes

➤ Drainage, both inside and outside the house

➤ The presence of leaks and drips at fixtures and pipes

➤ The water heater

➤ Any appliances, including the dishwasher, clothes washer, and hot water heater

➤ The condition of the fixtures, including the finish and age

If bottled water sales are any indication, more than a few of us think about the source and soundness of our water but don't go to the trouble of testing it. Without doing so, we can only guess about its purity, although bottled water has to pass certain minimum standards according to law.

Does Clear Mean Clean?

Unless you have Perrier delivered by the tankerload, the source of water for your home will be provided either by your local municipality or by your own well. Either source can have quality problems, including these:

➤ Bacterial contamination

➤ Varying amounts of lead

➤ Taste, color, and odor issues

➤ Water that leaves stains on clothes or fixtures

Bacterial contamination is more of a possibility with private water sources than municipal sources. The latter can become contaminated during extreme flooding or breaks in the water mains. A private water source should be tested yearly for *coliform* bacteria. Other tests for nitrate, sulfate, chloride, iron, lead, and hardness should be done according to recommendations by your local water department.

If you have any questions about the quality of well water or other privately supplied water, you should have it tested.

Get the Lead Out!

Up until the late 1920s, lead pipes were still used, although not extensively, in residential plumbing systems in the United States. The water from your water treatment plant might be lead-free, but it can pick up some of this fun metal along the way. Remember, the word *plumber* derives from *plumbum*, which is Latin for "lead." Lead has been used since the Roman days, although its use has been almost entirely eliminated from the plumbing industry today.

An old plumbing system—either yours or the city main—can contain lead in the pipes, fittings, solder, and fixtures. It's generally not a major health issue, but if you have any concerns, you should test your water for lead. If it is present, you can take some precautions:

➤ Allow the water to run for a minute or so before consuming it (you're drawing off water that has been sitting in the pipes and has the highest lead content).

➤ Never use hot tap water for cooking or preparing baby formulas (hot water leaches more lead from the pipes).

➤ Install a water filter that removes lead.

Young children are the most susceptible to lead-related health issues.

Plumbing Perils

Call your local water department to get the scoop on testing your water for lead and other contaminants. Lead test kits are available at paint and hardware stores, but their accuracy has been questioned by the Environmental Protection Agency, which recommends other methods. The most accurate test will be done by a laboratory specializing in water quality testing.

Oh, the Pressure

One immediate finding during your inspection is a sense of your system's water pressure. It's simple: Turn on the tap at the bathtub and the bathroom lavatory, and then flush the toilet. If your water flow remains strong, it indicates that you have good water pressure and properly sized pipes. A seemingly inadequate flow of water could indicate either low pressure or undersized pipes that can't meet water demand.

An old system using galvanized pipes might be so corroded that the interior diameters of the pipes have shrunk and are restricting the water flow. There's nothing you can do about this except replace them (there's no angioplasty available for pipes). Sometimes the pipes might be okay, and you can increase the water flow with a new, larger-diameter service line, but there are no guarantees.

Pipe Dreams

Not sure if your pipes are lead, copper, brass, or galvanized? Use a magnet! It will stick to only ferrous metals, such as the galvanized steel. Lead will be a dull, gray color, and copper and brass can be polished up with a bit of steel wool.

Types of Pipes

A newer house will have uniform pipes throughout, either copper or plastic. An old house can have a little bit of everything: galvanized, copper, plastic, and maybe even a bit of lead. A mix of pipes indicates that repairs or additions have been done to the system. This is fine—after all, you're considering alterations to the plumbing as well—if the work was done to code and passed an inspection.

If you find any galvanized pipe in your house, it's likely that this is your home's original plumbing, and anything other than galvanized has been added on. Look carefully at these additions. Do they have the correct fittings? Are the fittings stuffed with epoxy or another kind of synthetic filler? Are any kind of weird, homemade clamps used to connect two sections of pipe? (Don't be surprised—some past homeowners might have been very enterprising.) Record any of these observations in your notebook.

Plastic Pipe

There's nothing wrong with plastic pipe if it's the right kind for the application it's serving and is installed correctly. As we discussed in Chapter 3, "Codes, Inspections, and Safety," CPVC can be used as water supply pipe, but it might not be allowed by your local code. If you run across it in during your inspection, you need to know the pertinent local regulations regarding its use.

Flexible PB pipe and certain production runs of ABS drain pipe, both once approved by plumbing codes, are huge red flags to watch out for during your inspection (see Chapter 3).

Down the Drain

Close the drains in each tub, sink, and lavatory (this is a bathroom sink), and fill them with a good amount of water. Then open the drains and observe how fast the water

drains out. A slow moving drain is a sign of anything from a small clog in the fixture's trap to
problems in the sewer line. The latter is likely to be
true if all the drains are slow. A visit from a company specializing in drain cleaning might solve the
problem or only expose a larger one. What if the
experts discover that your sewer line is cracked and
packed with roots, dirt, and clay, and needs to be
replaced? Individually clogged drains can be an indicator of an owner who doesn't stay on top of
home maintenance. If you're considering buying a
house from this kind of owner, be extra careful in
your scrutiny during your inspection.

Outside drainage is also an inspection issue. Damp
or wet spots in a yard suggest clogged or nonexist-
ent drain lines. Yard drains empty either into a
storm sewer or into a French drain in a low point
of the yard. Over time, the drain pipe can get
clogged; old clay pipes can crack and need replacement or repair. Be sure to do your inspection on a
dry day, preferably a day or so after a rain storm.

Plumbing Perils

It isn't exactly a plumbing concern, but overflowing or leaky
gutters, whose water doesn't go
into a downspout to be routed
away from the house, can soak
the ground near your foundation. Cumulatively, this isn't a
good idea. Neither is a clogged
downspout drain. The idea is to
get water away from the house,
not soak the ground around it.

Look for Leaks

Obvious leaks, such as dripping faucets or basement pipes with puddles under them,
will be evident. Others might not be. Look under all the sinks and lavatories, and
around the base of the toilets. If you live in a humid climate and you're inspecting
during the summer, you might be looking at condensation, at least on a toilet. When
you're inspecting the basement pipes, check for temporary patches such as these:

➤ Hose clamps

➤ Pipe clamps

➤ Putty packed around fittings

A temporary patch can last for years, but you shouldn't accept it when purchasing a
house. Any defect we've mentioned so far can be a negotiating point when you make
an offer on a property. If you find one of these in your own house, you'll know to repair it or live with it, but at least you'll be aware of its existence. Drips and leaks
waste water and often become bigger drips and leaks, which might require a repair
when it's least convenient. (When is a plumbing repair ever convenient?) Mark this
defect in your notebook, and plan to repair it sooner rather than later.

Meet Your Water Heater

A quick inspection of your water heater (See Chapter 23, "Hot Water Heaters," for more information.) will tell you its age which will be listed on the inspection tag. Look for any indication of sediment buildup (open the drain valve and empty out a bucketful of water). You won't be able to tell whether it's working efficiently or what the condition of its components is, but you can figure that a 15-year-old water heater might need to be replaced soon. Note the size of the tank. Is this sufficient for your needs? Even if it's a fairly new unit, it might be woefully inadequate for your family of five if it previously served only a single individual.

Get to Know Your Appliances

Inspectors usually don't comment on the condition of electrical or plumbing appliances. You should take the time to run the dishwasher, washing machine, and disposer. Major appliances have varying life expectancies. Write down the serial numbers, and call the manufacturer's toll-free repair lines to date these appliances if you don't know when they were installed. When I called one manufacturer regarding a dishwasher part, the representative said that our particular dishwasher had an effective life of just over 11 years and it might not be worth putting any additional money into it because it was, conveniently, just over 11 years old. The point is, once you have this kind of information, you can make financial decisions that make sense for you.

Pipe Dreams

If you don't know how old long your washing machine hoses have been in service, and if the machine is more than five-years-old, go ahead and replace them with a heavy-duty hose. It's a small investment and can avoid the damage that a broken hose can cause.

Look at the washing machine hoses. Do they look cracked and old? If one of these breaks, everything underneath and surrounding the washing machine can get flooded. Make a note in your notebook to replace the hoses.

The Fixtures

A new house will have uniform fixtures in acceptable, neutral colors (white usually), unless it was custom built. This might mean one or more bathrooms with fiberglass or acrylic shower/tub combinations, a tiled shower stall in a master bathroom, mandated 1.6-gallon low-flow toilets, and a double-bowl kitchen sink.

If the house is a few years old, take a close look at the condition of the fiberglass or acrylic tubs and showers. Too many scratches and stains might not be acceptable to you.

An old house can have any combination of fixture styles, from original toilets with wall-mounted tanks to the cheapest of the cheap metal shower stalls. Examine and note the following:

➤ Loose or missing caulking

➤ Rust stains

➤ Chips in the ceramic coating

➤ Color and style

Fixtures usually have a bead of caulk between their rims and a surrounding surface (walls, floors, and counters). Missing caulking, especially around a tub, is a signal that the surrounding surface could be damaged from water infiltration. Caulking and grout between tiles must be kept intact to form a proper seal. Poke around with the blade of a pocket knife to test the area behind the broken or incomplete caulk for softness. If the blade goes through, there has been water damage.

Fixtures and Rust

Rust stains suggests that the water has a high iron content. They also indicate that a homeowner hasn't been much of a housekeeper if the stains are severe. Streak stains under a faucet indicate a drip that has gone without repair, or one in the past that went on for too long. Either way, you should test the water for iron and other metals.

They've Seen Better Days

A ceramic coating over a steel or iron fixture can last indefinitely, but that doesn't mean that it won't chip when an iron frying pan falls against the edge of a kitchen sink. The problem with chips and breaks in the porcelain isn't just a matter of appearance, but also a matter of sanitation. The underlying metal cannot be kept as clean as the porcelain. Repair kits are available consisting of small quantities of special epoxy paint and an applicator brush, but in my experience they're not a long-term solution.

Function aside, fashion might be an issue for you. Harvest-gold fixtures aren't exactly a hot decorator item these days. Can you live with the color scheme in your prospective new home, or do you see replacement in the future? Replacing all of a bathroom's fixtures is a major job and expense, and you'll have to figure this in to your budget.

Plumbing Perils

When applying touch-up epoxy paint to ceramic coatings, it's best to use several thin layers instead of one thick layer. Let each application dry according to the directions and then reapply. Single, thick coats will sag and run, the sign of an amateur job.

Septic Inspections

With periodic pumping and cleaning, a septic system should do its job indefinitely. Older systems used steel tanks, and these can be a problem because they can rust out. Check to see whether your system has a modern concrete or fiberglass tank rather than steel. Get a past cleaning schedule from the owner of the property and the name of the contractor who did the pumping. The frequency of tank cleaning depends on the size of the tank and the size of the household using it, as well as their disposal practices.

Plumbing Perils

Old hot water systems often have asbestos wrapped around the pipes in exposed areas. If you're tempted to remove it and replace it with modern insulation, call your building department first. Some municipalities prohibit homeowners from removing asbestos, while others allow it. Be sure to get a list of safety regulations if you're permitted to remove asbestos.

If possible, try to obtain a set of original plans and repair records for the system. Wet areas or darker-colored grass in isolated areas suggests that the system is leaching. Any backup in the system, of course, is obviously a problem.

Hot Water Radiator Systems

Both steam and hot water heat involve pipes, fittings, and water—in other words, more plumbing. These are old systems and rarely are installed today because of the expense of the installation. Modern hot water systems no longer use large, cumbersome radiators. Older systems need regular maintenance, such as flushing the system (see Chapter 23) and should be looked at by a qualified technician at least every other year to keep them in top operating condition. Think of it as cheap insurance.

Examine each radiator, and look for water stains on the floor under the control valves and fittings. This could indicate an ongoing leak or a minor drip. If you're considering buying a particular house, ask the owner for copies of the system's maintenance records.

The Big Picture

Consider the overall plumbing picture, and ask yourself these questions:

➤ Do you have an adequate number of bathrooms in your house or the house you propose to purchase?

➤ Do the existing bathrooms have adequate ventilation?

➤ Does hot water get to the fixtures quickly?

➤ Is the tile and floor covering in each bathroom in good shape?

➤ Are the kitchen appliances in good shape?

➤ Is the kitchen sink large enough?

➤ Are any bathroom floors spongy or sagging?

➤ Are there outside faucets at the front and rear of the house?

➤ Is the overall system in acceptable condition for modern usage and, more specifically, for your usage?

The point of this chapter is to help you gather information and make rational decisions about your plumbing. You might decide that you really can live with the existing fixtures after all, but that you absolutely must replace the hot water heater. The running toilet can be repaired, but the 1940s faucet cannot. Review your notes, add up the estimated costs, and then set up a plan.

The Least You Need to Know

➤ A thorough inspection of your plumbing will give you a good starting point for any plumbing project, especially major ones.

➤ An inspection will tell you whether your system meets your standards or needs improvement.

➤ You don't have to be a plumber to recognize telltale signs of trouble, such as water stains or low water pressure.

➤ Older homes will have the greatest mix of pipes, fixtures, and faucets, and also will have the greatest potential for problems.

➤ Never assume that past plumbing work was done to code or was inspected; look at the entire system with equal scrutiny.

Chapter 5

Your Wish List

> ## In This Chapter
>
> ➤ Some major plumbing upgrades
>
> ➤ Fixture and faucet choices
>
> ➤ Tub and shower combinations
>
> ➤ Kitchen plumbing
>
> ➤ Other add-ons and upgrades

The two most expensive rooms to remodel or build (per square foot) are traditionally the bathroom and kitchen. Bathrooms involve plumbing and fixtures. Kitchens require a water supply for certain appliances such as dishwashers and refrigerators with ice makers. The sky's the limit with either of these rooms, and the plumbing fixtures are only the beginning.

It's easy to get carried away here. You can install a basic, chrome-plated, single-lever faucet available at any home improvement center, or you can spend well over a thousand dollars on a minor work of art, a faucet sculpture, if you will, in a satin nickel finish. Some homeowners choose to restore existing antique faucets, which means new parts and replating or polishing all the metal surfaces. Tubs and sinks can be refinished, too, or replaced with traditional porcelain-on-steel units or new fiberglass-and-acrylic tubs and surrounds.

Modern kitchens, which are huge compared to those in older homes, are slowly having smaller second sinks added to islands or counter workspace so that several people can prepare food at the same time. Huge triple-bowl main sinks, almost 5 feet in length, are available for the primary sink. It all becomes a question of your needs, the size of your budget, and your sense of style—the most expensive variable of all.

What's That Thingamajig?

An individual **shutoff valve** will turn off either the hot or cold water supply to a fixture. The shutoffs are installed for safety in the event of a leak in the pipe between the shutoff and the faucet or in the faucet itself. They also allow for individual faucet repairs without having to shut off the entire water supply to the house.

Pipe Dreams

Sometimes it's possible to remove an existing main water supply pipe from under a driveway by attaching one end to a winch and slowly pulling it out. The new pipe is attached to the other end of the old pipe and is pulled along with it, eliminating any digging.

The Biggest Wish of All

The home improvements we can see are often the first ones we do. For instance, a house might have an old, dated electrical system with only one or two receptacles in each room, but some homeowners will repaint or wallpaper first anyway. When the system is updated, tearing into the walls to run electrical cable might necessitate redoing the rooms(s) all over again because of the needed wall and ceiling repairs. The same is true with plumbing.

An existing plumbing system might be entirely serviceable, but old galvanized piping can have restricted water flow. Individual *shutoff valves* could be absent from various fixtures. An undersized main water supply pipe—the pipe that supplies water to your house from the city's main—also means restricted water flow. Your plumbing system itself needs to be examined and evaluated before you start replacing fixtures or adding them.

Where Water Meets House

A main water supply pipe or service line could be as small as a half inch in diameter—and possibly corroded, at that. Service lines are sized to serve a certain number of fixtures, so adding a second bathroom could mean a drop in available water flow or volume. There are two issues here: One is water pressure at the main, and the second is the volume of water, which is determined by pipe size. Water pressure is pretty much a constant thing you have to work around, but pipe sizing can be changed. Your main water supply pipe should be your first consideration before adding any new fixtures onto an old system.

As pipes close down due to corrosion, they lose the capacity to carry enough volume to deliver a desired amount of water at the taps. Some older homes have pipes that are too small and undersized for modern requirements and demands.

Your water meter, which is often near the street or sidewalk, is connected to the supply pipe. You should be able to trace the probable path of the pipe to your house. Replacing the pipe often means digging a trench from the meter to your house. The pipe might be running under a driveway or sidewalk, and you or your plumber must find either an alternative route or an alternative to digging in order to replace it.

Hot and Cold Water Routes

Lead poisoning, particularly in children, has been a hot issue for years. In yet another example of the media trumping science, the true danger of lead and the sources of exposure have been held hostage by sweeping scare tactics by special interest groups. Nevertheless, old plumbing can be a source of lead that can then be ingested when you drink or cook with water passing through your pipes. Plumbing-based sources of lead include the following:

➤ Both city pipes and your house lines

➤ Solder and flux

➤ Pipe fittings

➤ Faucets

➤ Some ceramic fixtures

Until the 1930s, some lead piping was still installed for residential use. Lead solder was used until the 1980s. Water—especially hot water—can leach some of this lead. Water-borne lead is not a significant source of lead poisoning in most cases, though, and can be mostly avoided by allowing the *first draw* water to run for at least a minute before consuming it. Water sitting in pipes and faucets holds more lead, so a minute or so of opening the faucet will clear off this sitting water.

Lead aside, old galvanized pipes can become restricted with age as the metal deteriorates. If your inspection warrants it, you should consider completely repiping your house and installing new copper pipe. Like most infrastructure work, it's expensive and a lot of work to replace plumbing that you can't see. There's no escaping the basics, however, and it's pointless to install spiffy new fixtures and faucets if the underlying system is lacking. If

What's That Thingamajig?

First draw refers to any water that *is* sitting in the pipes when you first turn on a faucet. Removing the first draw water might take a minute or more. In an old system, allowing the water to run first will remove most lead that might have accumulated in the water while it remained in the pipes.

Plumbing Perils

In addition to the plumbing expenses when repiping, you'll also have some wall plaster or drywall repair and painting to do as you open up walls and ceilings. After the patching is done, you might have to redo some existing tile work as well. These expenses must be taken into account when you budget your plumbing work.

you have budget issues, do the repiping first and put off the all your visual improvements for another year.

Bathroom Fantasies

A bathroom serves three main purposes:

1. Bathing
2. Grooming
3. Housing the toilet

A fourth purpose might be to soak in a tub and hide from the world, but that's another matter. At its most basic, your bathroom could have a concrete floor, an inexpensive fiberglass shower stall, and equally inexpensive fixtures. We've gotten beyond that, though, with some designer bathrooms featuring fireplaces, wine racks, and exercise areas. As one reader to a national advice columnist put it, who wants to be eating in a room that is basically a dressed-up latrine?

I'm not going to preach about bathroom improvement limits. In the United States, the only limits are your checkbook, your local building code, and the available technology. Build the bathroom you want, but consider how badly you *really* want those gold-plated faucets and marble floors.

What's That Thingamajig?

A **lavatory** is what is commonly referred to as the bathroom sink. A distinction is made between the two because a lavatory is defined as a basin for personal washing, while a sink is mainly for washing things such as dishes. Lavatories typically are made from vitreous china and are more delicate than sinks.

Fixed on Fixtures

Fixtures are distinguished from faucets and pipes. A faucet sits on top of a fixture while connecting to a pipe and drawing water from it. Bathroom fixtures include these:

➤ Tubs
➤ Showers
➤ Lavatories

Toilets and bidets are also fixtures, but will be covered in Chapter 11, "Say Hello to Your Toilet."

Your choice of fixtures will depend on these factors, among others:

➤ The primary users of the fixtures
➤ Space limitations
➤ Budget

Most people (parent people, anyway) will make a distinction between a bathroom used primarily by kids and one used by adults, the former often getting comparatively utilitarian fixtures. The same could be true for an extra full or partial bathroom in a basement as well. A master bathroom is often fancier than the other bathrooms in a home. There's also the issue of how many people will be using the bathroom at the same time. You'll probably want two lavatories in a master bathroom and, if two are showering together, a comfortable-sized shower area with dual shower heads.

Space is a big determinant when choosing fixtures. The new bathroom you're thinking of adding might have only enough room for a shower stall, but not a tub and shower. Lavatory sizes vary, and you might find yourself settling for something smaller than you expected.

Plumbing Perils

Measure your space carefully. A single-person shower, for instance, requires a minimum space of 32 × 32 inches to the wall studs. Bathtub and shower space must be carefully sized and framed so that the drywall or tile meets the fixture properly.

Kohler RiverBath—a jetted tub with room for two.

Finally, budgets can kill your fixture fantasies faster than anything else. If you have only so many dollars to spend, your reproduction pedestal sink with high-end faucets can suddenly become a bargain basement particle-board vanity with a fixture and faucet manufactured in a former Soviet satellite and with installation instructions written in an unpronounceable language.

Plumbing the Depths

Two basic types of faucets exist: washerless and compression. A washerless faucet can be either a single-handle or two-handle model (the handles turn the water on and off). This faucet controls the flow of water with a replaceable cartridge or a sequence of seals that allow water to flow when their holes or "ports" are lined up correctly. Washerless faucets are designed for long life. Compression faucets use washers, either rubber or composition disks with a hole in their centers that look something like flat donuts. The washer closes against a metal washer seat when the faucet handle is turned. Leaks occur when either the washer or the seat wears out. Turning the handle tighter only increases the damage.

Speaking of Faucets

There's no end to the variety, style, and expense of faucets. Moen, a major fixture company, allows you to order a custom faucet by choosing among a variety of faucet bodies, accent kits, finishes, and handle options. Some faucets even come with built-in thermometers so that you know the temperature of the water, although most of us can figure out pretty quickly if it's too hot. Scald-protection safety valves also are available for showers and can especially help protect small children and the elderly. These valves work by maintaining a constant water temperature even when the water pressure changes.

Kohler Fairfax lavatory faucet.

Faucets come in a variety of finishes, including these:

➤ Polished chrome

➤ Polished brass

➤ Polished nickel

56

➤ Satin brass or nickel

➤ Gold-plated

➤ White, black, and other colors

Plumbing the Depths

One inspiration for bathroom fixtures, oddly enough, has been the American penal system. Known for vandal-proof toilets and sinks made from thick stainless steel, so-called prison models now are offered to those who find these essentially indestructible fixtures just the ticket for a contemporary bathroom. To view a line of these products, contact the Acorn Engineering Company, Inc. at www.acorneng.com or 1-800-488-8999.

You Can Call Them Lavs

Bathroom lavatories come in four general types:

➤ Freestanding lavatories

➤ Lavatories that are set into a vanity or counter

➤ Vanity tops that incorporate a lavatory

➤ Wall-hanging models

Freestanding lavatories, such as pedestal sinks, generally work best in large bathrooms where there is plenty of storage and where you're not dependent on a vanity or cabinet around the lav. These lavs are more expensive, as a rule, than the other types.

Lavs that are set into a counter often are used because the cabinet is custom built to fit a certain space. This arrangement allows the homeowner to choose from a variety of counters, including tile, plastic laminate, and marble. When the cabinet is built and installed, the lav itself is installed in a precut hole in the counter.

Sinks are set into counters in one of three ways:

➤ An under-counter lav uses the edge of the hole cut into the counter as a rim.

➤ A self-rimming lav fits into the counter hole and has a rim that rests on the counter.

➤ One-piece vanity tops and bowls that fit seamlessly on top of the counter or vanity.

If you don't mind somewhat limited choices in counters, a one-piece top and bowl is a great way to go. The biggest advantages to this arrangement are that it's easy to clean and also seamless, which means no chance of water seepage between the edge of the sink and the counter. This is the simplest bathroom lav arrangement possible and is a great idea for the children's bathroom or a guest bathroom.

Lavs that sit on top of counters are known as above-counter basins. These are slowly working their way into some bathroom remodels and custom homes. Some people consider them to be very stylish, while others will view them as an inconvenience to use and clean because the outside of the bowl is exposed. Wall-hung lavs are great for small spaces such as half-baths.

Kohler Bateau Vessels lav—an example of an above-counter basin.

The plumbing to hook up your lav will be the same regardless of the type you choose or its price tag: hot and cold water in, waste water out. If you have plenty of space to fit your lav, then you have plenty of options when choosing a model and style.

Tub People vs. Shower People

Although they may have been a bit public for modern tastes, the Romans had the right idea when it came to bathing. They built large, luxurious baths with a plentiful supply of hot water. Nowadays, most full bathrooms come with a tub, but we have expanded on the concept and added jetted tubs for two, with multiple shower heads. The choice of materials and styles guarantees something to satisfy every taste:

➤ Fiberglass or acrylic tub/shower combinations, one-piece units with seamless walls made from the same material on three sides of the enclosure

➤ Heavy duty porcelain-on-steel tubs

➤ Modern versions of classic claw-foot tubs

➤ Refinished old tubs

➤ Shower walls and floors constructed with tile, marble, or other stone

There's a lot to be said for installing fiberglass or acrylic tub/shower units. They're easy to clean and maintain, they install quickly, and they are competitively priced with other systems. Acrylic is considered to be tougher than fiberglass and has less tendency to fade. Because the units are seamless, you don't have to worry about water leaking where the tub meets the surrounding walls.

What happens if you want an acrylic unit, but you don't have enough room to get it into your bathroom? You won't need to grab your electric saw because units are available that come in sections. Their look isn't as clean as the one-piece units, but they'll do the job. Buy the highest-quality unit available (they will be manufactured by the same companies that make high-end one-piece units).

There are steel tubs and then there are steel tubs. Thicker-gauge steel tubs retain water temperature longer than narrow-gauge steel. Thinner walled tubs won't allow you to soak and leave the world at bay for as long as a heavier tub.

Old Tub, New Tub

Some people absolutely swear by the comfort of older tubs, especially the claw-foot models from the turn of the century and earlier. These tubs were usually huge and made from heavy, cast iron coated with porcelain. Say what we will about the Victorians, they knew how to build bath tubs. Modern versions of claw-foot tubs are made for both restoration work and general installations.

You might have an older tub whose appearance and finish don't quite match its comfort level. Porcelain finishes can take only so much cleanser and dripping faucets over decades of use before they lose their luster and develop rust stains. Fortunately, the wonderful worlds of chemistry and industrial know-how have stepped in with various refinishing processes to bring these tubs and other fixtures back to life. When you employ these companies, a technician comes to your house and recoats the fixture with a high-tech finish. This is not a new coat of porcelain, and the results vary depending on the process used and the skill of the technician. If you're considering this option, keep these pointers in mind:

Pipe Dreams

You don't want your plumber messing around in your newly refinished tub, so have as much plumbing work as possible done before the refinishing. The refinishing company will tape off and protect the faucets and the tub spout. Discuss this with the refinishing company first so you'll know their preference.

➤ Get at least two or even three bids over the phone and a description of each company's refinishing process.

➤ Ask to look at some of the company's past work, preferably a fixture that was refinished some years back; this is a good test of durability.

➤ Inquire about the experience level of the technician.

Before the fixture is refinished, it should be scrupulously prepared and cleaned by the technician. The finishing coats can be pretty toxic-smelling, and you, your children, and your pets might want to spend the day out of the house while the work is being done.

Tub Walls and Shower Stalls

The traditional material to cover tub walls and shower stalls is tile, which sheds water well and protects the underlying wall. The tile will shed water forever, but its grout, the cement-like material that seals the gaps between tiles, is another story. Grout is partially porous and must be cleaned and resealed as often as once a year. It's also a nuisance to clean because grout is a mildew and mold magnet.

In some respects, tile is only as good as its maintenance and its underlying wall. You can control the maintenance, but how do you know what's underneath the tile? Depending on their age and the builder, tub and shower walls will be one of several materials:

➤ A thin bed of mortar or mud in steel mesh

➤ Plaster

➤ Drywall (plasterboard)

➤ Cement backerboard

Back in the real old days, tile was set in mortar and would last for decades and decades. You'll know how tough this system is if you ever have to tear apart a tile wall or floor set in cement. Plaster would be used around a freestanding tub if there was no contact with the wall, although as tubs got replaced, some homeowners butted a new tub against the plaster, with or without covering the wall with a water-shedding material.

Drywall is a term describing sheets of paper-coated, gypsum-based wall covering. It comes in various thickness and lengths and four foot widths so that it will easily nail to the existing spacing of wall studs and joists. Greenboard has a more water-resistant paper covering, but it isn't waterproof. It shouldn't be used as a backing for tile, but it often is, much to the regret of homeowners who find water seepage behind their tile. The only backing system for tile that works, short of a layer of mortar, is some form of cement board, such as Wonder Board or Hardibacker.

Cement board is a generic name for tile-backer systems. They are typically made from cement and fiberglass mesh, and attach to wall studs with either screws or nails, just like drywall. These backer systems also work for marble, Corian, or other stone finishes used around tubs or showers.

Kitchens Need Water, Too

Your kitchen might not be your biggest water user, but it makes special demands. You need water for cooking; for cleaning hands, food, and dishes; and for drinking. That seems straight forward enough, but answering these needs can get complicated. Do you buy a double-basin sink or a larger triple-basin model? If the kitchen is large enough to accommodate a second, smaller sink, should you install one so that two or three people can easily prepare food at once? You can drink from the tap, but what if you want a refrigerator with an ice and ice water dispenser? That's going to need a pipe running to it, too. And then there are appliances: Just about every new house has a dishwasher and a food *disposer*. All these are plumbing concerns and can be met in a variety of ways. You even have to consider whether you want to install a water filter at the sink!

What's That Thingamajig?

Disposer is the correct name for the sink-mounted appliance that grinds up food scraps, although *disposal* is often used. The food disposer was invented by John W. Hammes in 1927, who eventually started the In-Sink-Erator Manufacturing Company, a major producer of disposers.

Kohler PRO TaskCenter with ProMaster faucet. This kitchen sink should satisfy everybody.

Take the matter of replacing your current sink. You're safe going with the same size, but installing something larger can be a major operation depending on your cabinets and the amount of space available. You might have your heart set on that Kohler PRO

TaskCenter 60-inch double-basin sink with a built-in drain board, but you can't change the laws of geometry: You either have room for a larger sink, or you don't. Suddenly your small plumbing operation becomes major surgery.

Watering Holes

We use kitchens and bathrooms regularly and are willing to splurge some on fixtures and faucets. Other water needs must be met as well, including the laundry room, hot water heaters, scrub sinks, and outside water.

Fortunately, the world of fashion and high design haven't honed in on scrub sinks and water heaters yet. These tend to be mostly utilitarian and very much oriented toward a working household.

Washday Blues

Laundry rooms use to be housed in the basement, garage, or ground floors of houses, but that's changing. Some builders are now installing the laundry on the second floor, closer to the bedrooms, where most of the dirty laundry originates. This makes a certain logistical sense, but it poses some problems if these locations aren't built to accommodate a washer and dryer. You'll have to consider these issues as well if you decide to move your laundry to a second floor:

➤ Providing a *dedicated electrical circuit* for the washer and for the dryer

➤ Venting the dryer

➤ Providing a floor drain or pan under the washer in case of leaks

➤ Stabilizing the floor if the washing machine abruptly moves from an unbalanced load

Most second-floor laundry rooms are built near an existing bathroom so that they can share the water supply and tie into the drain-waste-vent system. Separate, dedicated electrical circuits must be run for each appliance. A dedicated circuit is used exclusively for one electrical load, usually a major appliance.

What's That Thingamajig?

A **dedicated circuit** is one that only supplies power to a limited load. A load is an appliance, light fixture, or any other device that uses electricity. Dedicated circuits usually serve heavy power demands and would become overloaded if any other device was running on them.

Don't Forget the Drain!

The damage from a burst washing machine hose may be covered by your insurance (check to be sure), but who wants to rebuild and replace damaged walls, ceilings, and floors from hundreds of gallons of water? Pans are made to fit under the washing machine and catch any light drips, but not major leaks.

A more expensive and reliable choice would be a floor drain or, as an alternative, the FLOODSAVER from AMI, Inc. (See Chapter 24, "Laundry Rooms.") A floor drain works best with a tile floor built over a tile backer system, not plywood or oriented strand board (OSB), which is typically used in new construction. This really leads up to the question: How likely is it that you'll have a leak that would require this kind of a back-up system? If you took a poll of your friends and family, you'll probably find very few burst washing machine hose stories (and probably few replaced hoses, either). Nevertheless, if you're planning a new laundry room, you have to weigh the possibilities and decide whether the odds are in your favor against a major leak occurring.

Plumbing Perils

A broken washing machine hose can leak up to 500 gallons of water an hour. A standard hose should be checked regularly for bulges and pinhole leaks, and should be replaced every five years. An extended life hose (about $30) will stand up to water pressure even longer.

Getting in Hot Water

Hot water when we want it is one of the great comforts of civilization. Few of us want to start the day with a cold shower. Hot water heaters or tanks demand a water supply, of course, but not a connection to the DWV system. A gas water heater will have its own vent, and an electric water heater doesn't need a vent. All tanks come with a temperature-pressure release valve at the upper side of the tank and a drain valve at the lower portion of the tank.

Hot water heaters can be located anywhere along the cold water supply, but they often are close to the entry point of your service or supply line, which carries water in from your water provider outside your home. The gas companies and electric utilities each favor their model of hot water heaters (see Chapter 23, "Hot Water Heaters"), but check both types before you decide to install or convert to one energy form or the other. An electric water heater requires its own dedicated circuit, and a gas model requires gas piping, a task best left to a professional plumber.

Water heaters have an average life of approximately 12 to 15 years. If yours is up for replacement, consider whether you need a larger size tank. You may be running short of hot water because of the condition of your old tank or because it's inadequate to meet the needs of a growing family.

Working Sinks

After a day of mucking around in the yard, or mucking around with your plumbing, you'll want to scrub up in a real sink instead of using the garden hose. A good basic scrub sink, usually found in a laundry or utility room, will come in handy for really dirty hand washing and also washing out latex paint brushes and roller covers. You don't have to go with the standard, utilitarian fiberglass laundry sink, either. An inexpensive steel kitchen sink installed in a utility room counter makes a handsome scrub sink with plenty of room for potting plants or scrubbing under fingernails. If your local building code allows it, you can even install a scrub sink in a detached structure such as a garage or garden shed, using a special dry well for drainage.

What's That Thingamajig?

A **sillcock** is simply an outdoor faucet. It's constructed to hold up to the weather and is a pretty basic affair, unlike a more decorative indoor faucet. Sometimes the term hose bib is used synonymously with sillcock, but some manufacturers and plumbers say this is incorrect and that a hose bib is a separate type of faucet used for a clothes washer, for example.

The Great Outdoors

Typically, a new house will have a *sillcock* or outdoor faucet at both the front and the rear. Some older homes will have only one sillcock (also called a hose bib). A 100-foot hose will take care of most watering needs even with a single hose bib, but you might want the convenience of running more than one hose at a time. And you don't need to be limited to attaching your sillcock to the house itself. You can install one anywhere you can run pipe, which means just about anywhere in your yard.

Sprinkler systems made with flexible polyethylene (PE) or PVC pipe make watering your yard as simple as setting a timer. The pipe will have to be buried, of course, and the layout of it and the sprinkler heads will have to be planned so that your yard is properly watered. In cold climates, these systems need to be drained for the winter to avoid damage to the pipes and fittings.

Where Will You Pipe Today?

The human body needs a minimum of 64 ounces of fluid—preferably plain water—for good health. We also need water for cooking, maintaining our personal hygiene, keeping the azaleas alive, and washing our cars. Your wish list can extend and enhance your plumbing system to whatever your budget and your time will allow. Your costs will be roughly divided into two major areas:

1. The rough-in work, which includes installing all the water pipes, vents, and drain lines

2. The finish work, which includes installing all the fixtures, faucets, water appliances, tub and shower surrounds, tile or plastic laminate work, vanities and counters, and wall repair

There's no escaping the infrastructure costs. If you want water in your new bathroom, you have to run the necessary pipes, vents, and drains. You usually can control the fixture costs by choosing inexpensive models instead of top-of-the-line ones. The whole point of a wish list is to start you thinking of your own plumbing needs and how you can implement them. By now, you have a good idea of the basics of your plumbing system, its strong points, and the areas that it's lacking. In the chapters ahead, you learn how to make the best of this system and how to improve it to serve you best.

The Least You Need to Know

➤ The first plumbing change to consider is replacing or substantially upgrading an old system if it's in poor condition.

➤ It's easy to get carried away remodeling a bathroom; look at the long-term expense and benefits before you start the job.

➤ Kitchen plumbing can get complicated when you start figuring in dishwashers, extra sinks, disposers, and ice makers.

➤ Plumbing costs are split between the expense of installing pipes, drains, and vents, and the expense of the fixtures. You can always lower your costs by choosing modest or inexpensive fixtures.

Do-It-Yourself or Hire It Out?

There's a notion that our pioneering foreparents did everything from making their own shoes to mining iron ore and making nails in the family steel mill. To some extent this was true, but economies develop because some people do some tasks better than others. As a result, we trade skills and services for those that others do better. Carpenters, farriers (who made and fitted horseshoes), and millers were sought-after practitioners of their trades just like plumbers are today.

Sure, you could learn and eventually carry out many of the practices and trades that affect your daily life (I'd skip attempts at self-surgery, though). You'll gain some independence and satisfaction from learning new skills, but you'll face a learning curve as well as lost leisure time.

You should have a basic working knowledge of your plumbing system (and other household systems as well) and know enough to keep the drains clear, the toilets functioning, and the faucets from dripping. "Knowing thyself" will tell you how much further you want to go. If you must hire a plumber, you'll need enough knowledge to discuss the problem and the job intelligently, and write a contract or letter of agreement when necessary. This chapter runs you through the basics and helps you decide how much plumbing you want to do for yourself.

Call Me Plumber

Most of us can learn and familiarize ourselves with the basics of plumbing. This doesn't mean we'll have the skills or desire to install the entire system for a new house, but we can know enough to deal with the day-to-day problems and maintenance. How much can you expect to know as an amateur? For a start, you should know these points:

➤ Where your main shutoff valve is located

➤ How to use a plunger to clear drains and clogged toilets

➤ Emergency repairs for leaking pipes

➤ How to stop drips

➤ How to adjust a toilet tank float arm

➤ How to replace your washing machine hose

➤ The proper way to winterize your pipes

➤ When to drain your water heater

Everyone should know how to shut off the water in a house. If a pipe breaks as the result of a freeze, you can end up with a flooded basement or crawl space, and neither is a very pretty sight. Shutting off the water means that you won't have to call a plumber out in the middle of the night at middle-of-the-night rates. You'll be able to deal with the problem during normal business hours, when plumbing suppliers are open and plumbers can work in the daylight.

Pipe Dreams

After you've located your main shutoff valve, test it to be sure that it moves easily. You might have to apply some penetrating lubricant. Locate and test all your individual fixture shutoff valves as well. Then you'll be sure that they're in good working order if you need them later.

Plungers Are Your Friends, Too

There's a reason that a plunger is called a plumber's friend. A plunger solves all kinds of stoppage problems and keeps them from becoming bigger problems. Waste pipes are designed to handle liquids, not solids such as hair, food, dental floss, or cooking grease (disposers aren't even meant to take excessive grease). Toilets can absorb only so much waste and bathroom tissue. Overburden the system, and you have to relieve it of its burden by clearing the clog.

Be sure that you have at least one good plunger in your house. Teach your kids how to use it, too (you can never time a clog). The next step is using a snake or a closet auger, but at a minimum you should be able to use a plunger.

Plunger rules are simple:

➤ In sinks and tubs, stuff a rag or washcloth in the overflow hole or in the drain hole of an adjoining sink.

➤ Remove the stopper from the drain.

➤ Rub a thin layer of petroleum jelly on the rim of the plunger for better adhesion to the fixture.

➤ Push down and pull up vigorously until the clog is cleared.

➤ Run hot water down the drain to clear away any remaining obstruction.

➤ In toilets, use a plunger with a flange (a second, smaller cup within the larger cup).

Plungers are as low-tech as you can get, but they get the job done. You'll never have to worry about your plunger becoming outdated or needing a fix or a software patch because of a plunger virus. See Chapter 10, "Clearing the Clogs," for more help on clearing clogs from your drains.

Leaks Are Never Convenient

Pipes are always under pressure as long as they're connected to a water supply. A plumbing system is only as good as its materials and its installation, but age gets the best of all of us. Given enough time, some leaks can occur. Galvanized pipe is often the most vulnerable, but any material can leak. You might not want to cut out and replace a section of pipe, but you should know how to stem a leak until a plumber can do the repair.

Even if it's Thanksgiving afternoon and every hardware store is closed, you often can find materials around your house or garage to do an emergency patch on a leaky pipe. If nothing else, after shutting off the water, you can cut up a garden hose, split the sections lengthwise so that they can fit around the errant pipe, and secure them with wire twisted around like a tourniquet. Place a small wedge of wood against the hose first so that the wire doesn't cut the hose. You can then partially open the main shutoff valve, which will cut down on your water pressure but still give you some flow at your faucets. It's not a great repair, but it should

Plumbing Perils

Never use a plunger if you've poured a chemical drain cleaner into a sink or tub. These chemicals are caustic and can injure you if your plunging splashes them on your skin or into your eyes.

Pipe Dreams

Keep some pipe repair materials on hand for those unexpected emergencies, especially if you have an old plumbing system. Hose clamps, pipe clamps, epoxy putty, and even a scrap rubber inner tube can come in handy until a full repair can be done.

do the trick until a plumber shows up the following day. See Chapter 12, "Easy Leak Repairs," for more on this subject.

Float Arms

Your toilet float arm (see Chapter 11, "Say Hello to Your Toilet") opens the intake valve or ballcock, allowing water into the tank and shutting the valve when the tank is full. Sometimes the float arm needs a slight adjustment if there is too much or too little water in the tank. Bending a metal arm, or adjusting a plastic arm, is a simple bit of maintenance that does not require a plumber (see Chapter 11).

Drip, Drip, Drip

Water will flow where it has the least resistance, which means some kind of an opening. A hole in a pipe or a deteriorated valve is the plumbing equivalent of an "Exit Here" sign. Besides being annoying, drips are a waste of water and will only get worse as the valve or faucet continue to wear down. It's well within the skill levels of most homeowners to tighten compression fittings, replace washers, and replace the various parts of a washerless faucet. (See Chapter 13, "Start with a Faucet.")

Sink traps can also leak from deterioration and from loose fittings. All hardware stores sell replacement traps and couplings that can be installed without any special tools.

Washing Machine Hoses

Replacing the washing machine hoses falls into the category of good ideas that never get done. Preventative maintenance isn't our mainstay, and we usually pay for not doing it when the furnace stops running, the car wheezes to a stop, or we have a plumbing problem that we could have avoided. You should examine a standard washing machine hose after five years and consider replacing it. An extended-life hose (about $30) will last longer than a standard hose. Replacement is simple:

➤ Shut off the water at the hose bib behind the washing machine.

➤ Remove the hoses and let them drain into a bucket.

➤ Attach and secure the new hoses.

➤ Turn on the water and check for leaks (tighten hoses if drips appear).

If you move into a house with an older washing machine, it's always a good idea to replace the hoses.

Frozen Pipes Are a Bust

Hawaii residents can skip this section, but anyone living in an area that experiences freezing and subfreezing temperatures should know about winterizing their pipes. Any water pipe passing through an unheated space, including walls without insulation and crawl spaces, is subject to freezing. When water freezes, it expands; your pipes will expand, too, as ice pushes against them and possibly forces a break in the pipe. You don't want to deal with that break once the pipes thaw and the water starts flowing again.

The simplest measures to protect pipes from freezing include these:

➤ Wrap the pipes with insulation.

➤ Use electric heating tape or cables.

➤ Allow water to drip out of your faucets during extremely cold weather.

Insulation is cheap and effective and should be wrapped around all exposed pipes. Some pipe insulation is sold in precut foam sections that conveniently fit around water pipes. For a homeowner, it's just a matter of scrunching inside a crawlspace and slipping the material onto the pipes. Electric heating tape or cable cannot be used under insulation, and some manufacturers do not recommend it for plastic pipe. If the power goes off, this approach is useless. The material is easy to install, however: It simply wraps around the exposed pipe and is plugged into a receptacle.

Moving water isn't as likely to freeze as sitting water, so opening your faucets enough to produce a drip is sometimes enough to prevent frozen pipes during extreme cold snaps. I don't advise depending strictly on this approach, however. After all, even Niagara Falls freezes over once in a while.

Plumbing the Depths

Some years ago, I read of a small town in New England that was expecting a severe cold snap. The residents had been advised to keep their faucets running at a drip level to prevent frozen pipes. Within a few days, the town water tower had been completely drained!

If You Must Thaw ...

Frozen pipes should be thawed gradually with a heat gun or a hair dryer. Avoid using a propane torch or an electric space heater! Every time we have a freeze here in Seattle, someone starts a fire in an unheated crawl space by thawing frozen pipes with a torch. When using a hair dryer or heat gun, keep the following in mind:

➤ Be sure that the heating source is grounded.

➤ Don't touch the pipe with the end of the heat gun or hair dryer.

➤ Move the gun or dryer back and forth across the pipe to thaw a broader area.

You also can wrap towels drenched in hot water around the pipes, but this is a slower process and requires changing the towels as they cool.

Pipe Dreams

You can avoid most sediment problems by draining a few quarts of water out of your tank every month. Simply put a bucket under your drain valve near the bottom of the tank and open it up (see Chapter 23, "Hot Water Heaters"). This water will carry sediment hanging around the bottom of your tank waiting to cause problems later.

Caring for Your Water Heater

According to Larry and Suzanne Weingarten, authors of the book *Water Heater Workbook*, a water heater can be made to last from 20 to 30 years (well beyond the usual life of 12 to 15 years) with regular maintenance. You might not want to go quite as far as the Weingartens advocate, but a homeowner can perform routine maintenance to either a gas or an electric water heater without calling a plumber.

One of the biggest enemies to a water heater is sediment, and that's particularly a problem with hard water. Suspended solids—especially calcium carbonate—precipitate out of the hot water and settle on the bottom of the tank. In electric water heaters, excessive sediment can cover the lower heating element and eventually cause it to burn out. Draining some water out every month will avoid problems later.

Other maintenance, including inspection of the thermostat, burner, and relief valve, and venting (on gas models) should be made by a qualified technician.

Call Me a Plumber, Quick!

Each of us has a cut-off line when it comes to home maintenance and repairs. I've known people who built entire additions almost single-handedly, and others who wouldn't clean their gutters on a one-story house. Some plumbing chores should be done by a plumber unless you intend to become a serious student of the trade. Otherwise, you need to know when to call in a professional.

In some cases, your local building code might make the decisions for you depending on what plumbing work they allow a homeowner to do and what work must be done by a licensed plumber. Some jobs that are often best left to a plumber include these:

➤ Replacing an existing plumbing system

➤ Installing and testing a new water heater

➤ Doing any gas pipe work

➤ Installing a service line

➤ Cleaning a major sewer line

➤ Doing major emergency repairs

You can probably add more jobs to this list depending on your skill level. Other factors will enter into your decision to call a plumbing contractor: You must consider your budget, time availability, and family concerns. As a rule, trying to watch the toddler while installing a new kitchen faucet isn't a winning combination.

New from Bottom to Top

Replacing your old galvanized pipes with new copper or plastic is a major undertaking. A plumber or two-person crew that does a lot of remodeling work can get the job done quickly and efficiently, and can often keep water running to some of your fixtures during the course of the job. Try this yourself, and you might not have the water back on for days or weeks. This is a time to put aside your need for independence and do-everything-yourself attitude and hire a contractor.

Installing Water Heaters

Some books suggest that you install your own electric water heater, but I'm not a big advocate of it. A water heater is heavy and cumbersome. Besides installing the new water heater, you must remove the old one and have it hauled away. You'll also need a permit for both the electrical and the plumbing work. If you run into any installation problems, you could go without hot water until you solve them. There's nothing like a cold shower or two to provide a reality check. Leave this one to a plumber (and always leave a gas water heater to a plumber as well).

Don't Mess with Gas Piping

Natural gas can leak just like water, but with one major difference: It can ignite and explode as well. Enough said? Local building departments frown on do-it-yourselfers messing with their gas lines and probably won't allow it anyway.

Main Water Supply Pipe—Just Dig

Your water supply pipe is your main link to the municipal water system. It's analogous to the electrical service lines coming into your house from the utility pole. It's one thing to alter your plumbing system inside your house, but first you need to ensure that you have a solid and correctly installed connection to the city main. Because this is a critical link, I suggest that you have a plumber install it. At least you'll have running water while you work on the plumbing. You can save some money by digging the trench between the *curb stop* and your house.

What's That Thingamajig?

A **curb stop** is a valve that is situated between your water meter and the city water supply. It's owned by the utility and controls the water going into your house. This is the valve that you turn off if you get a leak between your house and the valve itself, and it's usually at or near the city curb.

Obliterating Sewer Gunk

A major sewer blockage can mean tree roots or worse. Cleaning the sewer line requires a large, motorized snake that can damage the line in inexperienced hands. Companies specialize in this work, and it should be left to them. They've seen it all before and won't be intimidated by large blockages.

It Costs How Much?

You can measure costs several ways. One is the actual dollar amount staring you in the face on a plumber's estimate sheet. Another is time, which leads to several questions:

➤ How long can I put off doing this work until it becomes critical?

➤ Would I be better off working extra hours and paying a plumber rather than doing this work myself?

➤ Is it worth it to me to get the work done faster by a professional instead of my plodding through it?

A $50- to $75-dollar-an-hour plumber might be a real bargain if your leaking toilet can get pulled, the wax seal replaced, and the toilet reinstalled while you're at work.

Pipe Dreams

It can cost thousands of dollars to replace a plumbing system, even if you keep some of the old fixtures. This is a major upgrade that will make your life more comfortable and convenient. It also will increase the value of your home, although it's difficult to provide an exact return on investment. It's easier to break it down into parts. For example, new or upgraded bathrooms return much of their cost—and sometimes they return more.

It's a terrific bargain if you have only one toilet and *must* get the work done. We pay for goods and services every day based on our needs, wants, budgets, and time restraints. You'll have to decide where plumbing repairs rate on your "To Do" list.

Compare this to buying a new sports utility vehicle that you want but don't particularly need. You'll be paying a lot of money for a depreciating asset. Suddenly new plumbing can look like a bargain.

Dealing With Plumbers

When you hire a plumber, you each have your respective responsibilities and expectations. You need to clearly communicate what you want done and the time frame in which it must be completed. The contractor must be equally clear in stating the work as he or she understands it from your plans, the cost for labor and materials, and a reasonable completion date. Any changes by either party must be negotiated.

This could be a new experience for you. You'll find this stranger and perhaps an assistant wandering around your house in work boots, possibly punching holes in the walls and shutting off your water from time to time. Who are these guys/gals and how do you deal with them? Suddenly you're an employer of sorts hoping that these new employees are going to work out before you write them a check.

You need to be knowledgeable about the work (that's why you're reading this book), not only about your plumbing system but also about contractors. A contract, regardless of whether it's oral or written, is a legally binding agreement. You need to know your rights, the contractor's rights, the bidding process, and payment schedules. Then there are the intangibles, your personal reactions to individual plumbers bidding the job. If red flags start metaphorically popping up in front of your eyes, you should look for another plumber.

Some plumbers specialize in repairs, while others do mostly new construction. Be sure that whomever you hire is geared up to do the type of job that you want done. Your first course of action is to *find* a plumber.

Plumbing Perils

In more prosperous times such as those we're experiencing now, contractors can be scheduled for weeks and months in advance. Even getting an estimate can be difficult, especially during the summer months. Plan ahead as far as possible to schedule your work.

Contracting for a Contractor

The best way to find a plumber is to ask other homeowners which ones they have found to be satisfactory—and which ones they haven't! You might not be able to schedule your first choice, but you'll know who to avoid for being less than satisfactory in their work or billing practices. Ask your neighbors, friends, and family for recommendations. Anyone with an old house has probably hired a plumber at one time or another.

There is no absolute guarantee that a contractor full of customer praise will give you the same results in your own home, but there's a good chance that you'll be satisfied with the results. Most small contractors survive on referrals, and they'll want yours as well.

Three-in-One

A legitimate contractor will be licensed, bonded, and insured. Without all three of these qualifications, you're putting yourself and your home at risk. If a cash-only, unlicensed, we-don't-need-no-stinkin'-contract plumber works on your house and incorrectly connects pipes or drain lines, you have very limited recourse when pursuing compensation for damages. You could even be subject to a claim if the plumber or an employee slips and is injured on your property. When a licensed plumber causes a problem, you have some legal that assurances the problem can eventually be rectified.

Licenses Issued Here

A plumber should be licensed, bonded, and insured in accordance with local and state laws. These requirements are fairly standard across the country. They protect you and the contractor from each other if problems arise.

A license is simply permission by a governing authority to carry on a specific business. It shows that a contractor is registered, often with both the city and the state, and has met certain standards to be licensed. At that point, a contractor can hang out a shingle and say, "This is me, plumber for hire." It also means that the local government has collected a registration fee and will be collecting taxes from the licensee.

Two major requirements normally must be met before a contractor's license will be issued:

1. The individual must be bonded.
2. The business must be insured.

Even if your state doesn't require one or both of these requirements, you should! Some local laws are less stringent than others, but that doesn't mean you can't hold your contractor to higher standards.

Plumbing Perils

Check that your plumber's license, bond, and insurance are current. All three are renewed on a yearly basis. If you have any questions at all, call your city or state department of licensing and do a credentials check. You don't want any problems from hiring an unlicensed individual.

The Name's Bond, Surety Bond

A contractor's bond, or surety bond, is required in many states before a contractor will be issued a license to operate. The bond helps guarantee that a contractor will perform according to the terms of a contract. I suppose it's not much different in principle from a jail bond, which is an attempt at guaranteeing a defendant's appearance in court, but with a more wholesome connotation.

A bond is registered with some governing authority in one of two ways:

➤ A contractor can establish a special account with a cash deposit equivalent to the amount of the bond.

➤ A bonding company can be engaged for a fee.

The amount of the bond varies from state to state. Here in Washington, for instance, the bonding rates are relatively small. A general contractor has to post

What's That Thingamajig?

Webster's Dictionary defines **surety** as "a guarantee, assurance, certainty" and dates the origins of the word back to the fourteenth century. The term *surety bond* dates back to the early twentieth century as "a bond guaranteeing performance of a contract or obligation."

only a $6,000 bond, while a specialty contractor or subcontractor—plumbers, electricians, painters, and the like—take out a $4,000 bond. If you are not satisfied with a contractor's work, you can put a claim in against the bond, although you're limited to its dollar amount unless you pursue other legal avenues for financial satisfaction.

Any claim against a contractor must be legitimate. You must prove that the work was not done to the specifications agreed to in your contract. Just as a bond gives you some leverage in the event of faulty work, a lien (rhymes with lean, appropriately enough) gives a contractor some protection against a customer's spurious claims. Sometimes called a mechanic's lien, this handy piece of legal work allows a contractor to file a claim against your home until your debt is satisfied.

Insurance Is a Must

Proof of insurance is usually a requirement for a contractor to obtain a license. It protects you if there's an accident or damage during the course of the work. Otherwise, you and your insurance policy might end up paying the price. Aside from a general liability policy, a contractor must cover any employees with government-mandated policies such as workers' compensation.

Plans and Specifications—Always!

For general repairs, you won't expect a plumber to submit highly detailed plans, although you should expect an estimated price. Repairing a burst pipe is fairly straightforward and doesn't lend itself to a lot of interpretation. Larger jobs are another matter.

You can't expect someone to bid a job without you specifying what you want done. You need to specify the type of pipe you want (plastic or copper), the fixtures, the faucets, and all finishes. Your plumber will tell you whether certain locations will work as you plan or whether you should reconsider your plans.

Details start creeping in when the scope of the job increases. Installing a new water heater could mean a different location than the one you're currently using. A complete updating of your plumbing system would have to be very detailed.

Allowing Substitutions

As remodeling bids come in and budgets get stretched, that imported marble counter often becomes plastic laminate, and the oak floor becomes vinyl. The same is true with plumbing work. Fixtures, faucets, and plumbing appliances come in a wide range of prices and quality. Sometimes your plumber can come up with an equivalent-model fixture at a lower price with no appreciable difference in quality or visual results. Your bids and specifications should allow for such substitutions once you have reviewed and approved of them.

Comparing Bids

A clear set of plans and specifications allows the bidding plumbers to all play by the same set of ground rules. This allows you to fairly compare their prices. You'll find as you put a job out to bid that each plumber might have a slightly different take on how to do the work and what materials to use. Keep these suggestions in mind as you scrutinize the bids so that you can adjust for specific differences in cost.

About Those Contracts

Some contractors—and homeowners—want a written contract for everything. This is unnecessary for small jobs, but there's no harm writing up a short letter of intent. You could say, for instance, "Contractor will supply all labor and materials for installing one new Mansfield bathroom toilet and lavatory for the sum of _____ dollars plus applicable tax. Homeowner will take care of any wall repair or patching." For that matter, your contractor might supply his own contract form for small jobs with a written description of the work and then ask for your signature to confirm your acceptance.

Larger jobs normally require a written contract. If your plumber is hesitant to provide one or sign yours, find someone else to do the work. No legitimate contractor will shy away from a valid contract.

Plumbing the Depths

Aside from describing the job and the materials to be installed, a contract also will state the terms of payment. Some contractors will want a percentage in advance depending on the size of the job. Some states limit this percentage. You also could be asked to pay for special order materials in advance. If the demand for prepayment seems unreasonable, find another contractor.

Change Orders

A change order is a modification to a contract. It can be initiated by either you or your contractor, but it must be agreed to by both. You might decide that you want a different faucet, for instance, or your plumber might run across unforeseen problems, such as rotted floor joist under a toilet. Usually, a change order means an increase in the price of the job, but not always. You might decide to eliminate some fixtures or go for less expensive ones, thus lowering your overall cost.

Ideally, the best change order is *no* change order. Change orders can delay a job and could cause your contractor to undo work completed under your original specifications to accommodate the requested change. No plan is perfect. Remodeling is a kind of a fluid experience. As it progresses, you see things that you might not have seen during the planning stages. You might decide on shower doors instead of a shower curtain, a more expensive toilet, or even different strainers for the kitchen sink. Every change alters the job and the cost.

A Deal's a Deal

Once you've agreed to the job and signed on it, you have to hold up your end of things, too. This means:

➤ Clearing out the rooms or areas where your plumber will be working.

➤ Keeping your children at a safe distance from the work activity.

➤ Controlling your pets.

➤ Providing access to your house with a key, or being home at the start of the work day.

➤ Understanding that your contractor and any crew will need access to a bathroom and somewhere to take their breaks.

➤ Paying your bill in a timely manner; small contractors are especially dependent on regular cash flow, and you shouldn't unnecessarily delay payment.

Plumbing Perils

Your plumber can rightfully put a clause in your contract that he or she will not assume any responsibility for damage to household items left in the way or not adequately protected on the work site. This includes anything hanging on a wall that could loosen and fall from hammering, sawing, or drilling through the wall.

Being a good customer is just as important as being a good contractor, all of whom have stories about the customer from hell.

Clean-Up and Wall Repair

Plumbers install and repair pipes and fixtures; they don't repair any walls, ceilings, or floors that get opened up in the process. Of course, they should supply adequate protection for floors and perform reasonable clean-up. You'll have to attend to any repairs to plaster, drywall, or floors unless other arrangements are made. Some plumbers will have another contractor follow up with these repairs for an additional fee.

The Least You Need to Know

➤ Don't expect to do every plumbing repair, but know how to do the basic ones.

➤ Everyone in your household should know where the main water shutoff is located.

➤ Hire only a licensed, bonded, and insured plumber to work on your home.

➤ Clear, written job specifications help both you and the bidding plumber to come up with an accurate, fair price for the job.

➤ Sign only contracts that you fully understand; if you have any questions, ask away.

➤ As a customer, understand your responsibilities to your contractor.

Part 2

Tools, Techniques, and Fixtures

Your plumbing work will call for some specialized tools that you won't use in other areas of house repair, but the investment is small and most of them should last a lifetime unless you lose them or loan them out. (Just think, your legacy to your great-great-grandchildren could end up being a set of pipe wrenches.) Plumbing also calls for an assortment of small parts such as washers, solder, and putty that are helpful to have on hand. Forget everything you learned about just-in-time inventory control, and keep some of these extra items in your tool box so that you're not running out to the hardware store every time you find a dripping faucet.

Your house is one huge web of fastening: Wood is nailed to more wood, wires are secured to electrical fixtures, and pipes are joined to other pipes. Each type of pipe (plastic, copper, and steel) calls for different fastening techniques. Chapter 8, "Pipes: Joining and Fitting," discusses each of them and their obvious differences from one another.

Outside of pipes and fittings, the next big material items in the plumbing world are fixtures and faucets. You have showrooms and catalogs full of choices here, and the sky's the limit when it comes to pricing. We'll review the different materials and products available so that you're better prepared to choose among them.

Chapter 7

Tools and Spare Parts

In This Chapter

➤ Plumbing's tool requirements

➤ Care and feeding of tools

➤ Buying vs. renting

➤ A minimum tool set

➤ Repair parts to keep around

At some point in the future, after our genes are mapped and scrutinized, an academic paper will be written confirming the existence of the tool gene—or, more accurately, the gene that dictates the relentless male desire to acquire tools. Political correctness aside, if you give the average American guy a clear choice between buying a new one-horsepower bench grinder for which he has no immediate use and paying the monthly property taxes, the grinder will win every time. This has been going on since our Neanderthal ancestors first started using stone tools around 70000 B.C.E., and as any tool department at your local home improvement center will attest, it's never going to end.

Plumbing requires some common tools, such as screwdrivers, that just about everyone has around the house, as well as tools specialized to the trade. You don't need to buy everything—some are more economical to rent—, but a decent set of basic tools is very affordable. You also don't need the very best tools.

Plumbing requires more hand tools than power tools, but you need some of the latter if you add on a bathroom or have to cut into walls or floors. Any tool can be dangerous in careless hands, so safety is paramount, especially when you're working with spinning drill bits or reciprocating saw blades. Before you start your plumbing work, read through this chapter and determine which tools you'll need so that you'll have them on hand. There's little more frustrating than trying to do a plumbing job with the wrong tools.

Plumbing Perils

Some professionals shy away from top-of-the-line tools because they have more opportunities to damage them on a job site or lose them—sometimes through theft—as they move around to different locations.

Pipe Dreams

Yard and estate sales can be good sources of used tools. Look for those in the best condition, especially hand tools. The cutting edges should be sharp, and the ends of the screwdriver blades should not be worn or rounded. Turn on electrical tools and look inside the vented section. You don't want to see any excessive sparking or smell a "burning" motor.

While you're acquiring tools, you should make sure that you have certain spare parts in your tool box as well. A couple pipe clamps can keep a late-night leak sealed up until you can work on it at a more civilized hour. We'll discuss some of these parts as well.

Tool Basics

When it comes to bargains, hand tools are at the top of the list. A $10 hammer will last for years, perform a wide range of jobs, and prove itself useful each and every time you use it. Clothes will shrink, expensive dishes will break, and software will become quickly outdated, but a hammer will carry on. The same is true with screwdrivers, pliers, and wrenches, all tools used for plumbing work.

There are a few universal tool rules:

1. You get what you pay for.
2. Abuse it and lose it.
3. Tools are easy to misplace.

You don't need the very best tools, but you're not doing yourself any favors by buying the cheapest. Every hardware store has a bucket of bargain screwdrivers near the cash register whose blade tips are almost guaranteed to live a short life before they become rounded off and useless. A better screwdriver will have a shaft made from a harder metal, and the tip will keep its shape instead of wearing down like the bargain version.

Tool abuse includes using screwdrivers as pry bars and pushing an electric saw with a dull blade to cut through framing lumber while listening to the motor strain. These are great ways to break tools or burn out brushes . If you use your tools for their designed functions, keep them clean and dry, and pack them away at the end of the job, they'll last indefinitely.

Hand Tools

Hand tools for plumbing fall into several categories. There are tools that cut, tighten and loosen, grasp, measure, and clear away obstructions. The following list includes the essential tools for the home plumber:

➤ Bulb-style plunger

➤ Closet auger

➤ Plumber's snake

➤ Standard (slotted) screwdriver

➤ Phillips screwdriver

➤ Adjustable wrench

➤ Slip-joint pliers

➤ Channel-joint pliers (also called channel locks or arc-joint pliers)

➤ Set of Allen wrenches

➤ A measuring tape

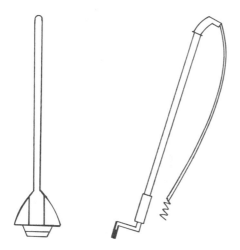

A plunger—the plumber's friend (left). Closet auger—an essential hand tool (right).

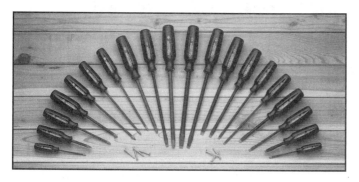

Standard and Phillips screwdrivers.

Craftsman Tools.

Adjustable wrenches.

Craftsman Tools

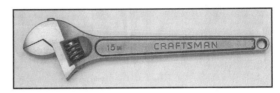

Slip-joint pliers.

Craftsman Tools

Channel-joint pliers.

Craftsman Tools

These tools will get you through most clogged drains and toilets, minor adjustments, and replacement of faucet parts. A bulb-style plunger (with a central flange for use in sinks) can be converted to a standard plunger by folding the bulb up and into the surrounding cup. A closet auger and a plumber's snake will clear most toilet and drain pipe blockages, respectively.

A convenient screwdriver is one that comes with several tips, both Phillips and standard, and sometimes magnetized. A good screwdriver tip also can be quickly installed and removed from the end of the shaft. An adjustable wrench (sometimes referred to as a crescent wrench) has jaws that open by turning a knurl, allowing it to tighten or loosen a variety of pipe and faucet fittings. A 10-inch adjustable wrench will tackle all kinds of plumbing repairs.

Slip-joint pliers are an all-purpose grabbing tool with two gripping positions. Its jaws are serrated and can scratch or mar finished surfaces, unlike an adjustable wrench, which has smooth jaws. Channel-joint pliers

Plumbing Perils

Be sure that your screwdriver tip is the right size for the screw you're removing. A tip too small can slip out and deform the head of the screw, making it harder to remove.

are kind of a marriage between an adjustable wrench and a pair of slip-joint pliers, except that they're not made for applying excessive torque to the item being tightened. They really offer quite a range of adjustment and are a very versatile plumbing tool. Channel-joint pliers have serrated jaws as well, so they must be used carefully around faucet fittings.

Old plumbing faucets won't require L-shaped Allen wrenches with their hexagonal heads, but some new faucets will. An Allen screw (also called a set screw) has a hexagonal or six-sided socket as a head instead of the more familiar slotted or Phillips heads. Allen screws are used to secure some faucet handles and tub spouts.

With these tools and a few parts, you will be able to take care of your occasional clog, drip, or leak. For bigger or more complicated jobs, you'll need some more specialized tools.

Pipe Dreams

Before using a pair of slip-joint pliers or channel-joint pliers on a faucet, wrap the fitting a couple times with electrical tape or a rag to protect the finish. You also can use duct tape, but that tends to leave more residual adhesive on the fitting than electrical tape will.

Tools You've Never Seen Before

Every trade has tools and procedures specific to its needs, and plumbing isn't any different. A common adjustable wrench, for instance, can be used when working on a bicycle, installing an electrical service, or repairing a drain trap. A basin wrench, on the other hand, is pretty limited to removing barely accessible faucet nuts from behind sinks.

The following hand tools will be needed for more complicated plumbing jobs such as these:

➤ Cutting and installing pipe

➤ Repairing faucet seats

➤ Removing steel pipe

➤ Repairing toilet leaks

Be sure that you buy a tool whose size can do a multitude of jobs. A 10-inch pipe wrench will cost a few dollars less than a 14-inch wrench, but it's a false savings if it limits the jobs you can do. The following is a list of specialized plumbing tools:

➤ Pipe wrench

➤ Monkey wrench

➤ Spud wrench

Plumbing Perils

Wrenches are used only on nuts or threaded steel pipe—in other words, where two pieces are joined by tightening and turning. Copper pipe and plastic pipe are joined by solder and solvent welding, respectively, and wrenches never should be used on them.

➤ Valve-seat wrench

➤ Strap wrench

➤ Basin wrench

➤ Locking pliers (vise-grips)

➤ Hacksaw

➤ Hole saw

➤ Tube cutter

➤ Valve-seat dresser or reamer

➤ Pipe reamer

➤ Tube or pipe bender

➤ Propane torch

Slip-joint pliers and adjustable wrenches are appropriate for some repairs, but you'll need some serious wrenches for bigger plumbing jobs.

A World of Wrenches

Different pipes and fittings require different wrenches for tightening and loosening. A pipe wrench is used on steel or iron pipe, but not on chrome or other polished-metal fittings unless you've got absolutely nothing else available (in which case, you must be sure to wrap the metal with rags or tape). Pipe wrenches, if you can find some in good condition, are good tools to buy at yard sales.

Each tool does a specific job.

➤ **Pipe wrench**—A pipe wrench has one fixed serrated jaw and one serrated floating-hook jaw that is adjusted with a knurl, which is an adjusting wheel. Pipe wrenches are usually used in pairs, with one wrench on a section of pipe and the other on its fitting. The jaws tighten as the wrench is pulled against a pipe; too much pressure can damage a pipe.

➤ **Monkey wrench**—This type of wrench is very much like a pipe wrench, except that the jaws are not serrated and thus can be used against finished metal.

➤ **Valve-seat wrench**—This tool removes valve seats from compression faucets.

➤ **Strap wrench**—This wrench has a canvas strap instead of jaws. The strap won't damage finished metal.

➤ **Basin wrench**—This one-of-a-kind tool is used to loosen the nuts securing sink faucets. Its long reach is perfect for tight spaces behind bathroom lavs and kitchen sinks.

➤ **Locking pliers**—This tool's jaws are serrated for a tight grip. The pliers are designed to lock onto a fitting or pipe and stay attached independently and hands-free. An adjustment screw at the end of the handle allows for various adjustments of the jaws.

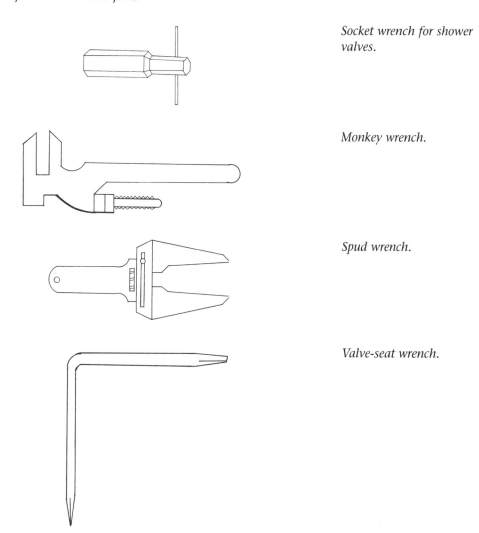

Socket wrench for shower valves.

Monkey wrench.

Spud wrench.

Valve-seat wrench.

Plumbing also requires tools that cut, clean, and manipulate pipe, bending it to desirable angles as dictated by a specific job.

Cutters and Hackers

Building materials come in standard lengths, widths, and thickness. They are then measured and cut to size on the job site. Carpenters do it with wood, and plumbers do it with pipe. You won't need a lot of cutting tools, but you will need those in the following list:

➤ **Hacksaw**—This saw is used to cut metal pipe and plastic pipe. The saw's fine, small teeth produce a clean, even cut.

➤ **Keyhole saw**—This saw is designed to cut through drywall, plaster, and lath. You might have to cut into your walls or ceiling to replace some pipes.

Hole saw.

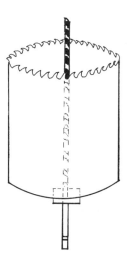

➤ **Tube-cutter**—Made specifically for cutting copper tubing as well as plastic pipe, tube-cutters come in different sizes, depending on the diameter of the pipe to be cut. Buy one with a built-in reamer to clean out the end of the cut pipe and re-move any burrs. A mini tube cutter is used for cutting in tight spaces.

➤ **Valve-seat dresser or reamer**—Sometimes called a valve-seat dresser, this tool cleans old, worn, and damaged valve seats in compression faucets where the valve seats cannot be removed.

➤ **Pipe reamer**—This cone-shaped tool cleans out any burrs or rough spots on the inside of a cut section of pipe. These obstructions can slightly obstruct water flow, but more importantly, they can facilitate calcium build-up, which can re-ally lead to blockage problems.

➤ **Tube or pipe bender**—This is a strong coiled spring used to bend copper tubing with minimum crimping of the copper.

➤ **Propane torch**—You'll need a torch to solder copper pipe connections, but be cautious around its open flame. An inexpensive model will do.

These are tools that can be bought or sometimes rented on an as-needed basis. There's no point in buying a new monkey wrench on a whim, because you'll never use it unless you have a plumbing project coming up.

Power Tools

Plumbing doesn't lend itself to a lot of power tool usage. Even if one existed, you wouldn't really need an electric wrench. However, you will need a few tools for cutting into walls and floors if have extensive plumbing to do:

➤ A drill (cordless or electric)

➤ A reciprocating saw

➤ A wet/dry shop vac

A cordless drill, which runs off a rechargeable battery, is wonderfully convenient because there's no power cord to pull around and get in the way. The drawback is the battery life, which varies with the tasks at hand. Drilling large holes through framing lumber will drain the battery life faster than drilling small holes in plywood.

A cordless-model drill usually costs more than a corded or electric model of equal size. Sales and close-outs at home improvement centers and tool stores can narrow the price gap. Two factors to look for when you buy a cordless drill are these:

1. Power (get at least a 12-volt model)

2. Charging times for the battery (the shorter, the better)

Drills are manufactured according to chuck size. The chuck holds the drill bit or other attachment, such as a grinding wheel or buffer pad. The larger the chuck, the bigger the drill motor because more power is required to drive larger drill bits and attachments. A manufacturer's line of drills includes $^3/_8$-inch and $^1/_2$-inch models.

A $^3/_8$ variable-speed, reversible electric drill will see you through many plumbing jobs. Prices usually start at around $60 and vary depending on the amperage of the motor (more amps are better) and the drill speed (revolutions per minute, or RPMs).

Pipe Dreams

Buy a second battery along with your cordless tool if you expect to be using it extensively. This way, you always have a battery charged and ready to go while the other battery is being charged. Professionals almost always have two batteries on hand and rotate them.

Cordless drill.

Makita USA

Every Little Bit Counts

Drill bits come in every shape and size for all types of jobs, from drilling through masonry to fine craft work. The most common bits you probably have seen are twist bits, which are sold both individually and in sets based on gradation. Twist bits are fine for small holes, but they're not much use for plumbing work.

Plumbers need to drill through wall studs and floor joists to run pipes, and this requires larger holes than a twist bit will provide. The most common bits for drilling larger holes are these:

➤ A spade bit

➤ An auger bit

➤ A power bore bit

These bits are commonly available both individually and in sets. Holes bigger than an inch and a half in diameter are often cut with hole saws. These are round steel drill attachments with saw teeth around the edge. The larger the hole, the easier it is for a drill to slip, especially when using a hole saw and working overhead on a joist. Be sure that you're standing steady and braced against the drill before pulling the trigger and activating it.

Reciprocating Saws

A reciprocating saw is an industrial version of an electric carving knife. It's great for cutting through walls, especially old plaster and lath construction, as well as cutting away old pipe. If you have only limited use for this saw, it might be better to rent (see the section "Rent or Buy?" later in this chapter) than purchase. This saw's blade moves fast—you might want to practice cutting on some old boards first to get a feel for the tool and its movements.

Reciprocating saw.

Makita USA

Sucking Up

A wet/dry vac is the perfect clean-up tool for home plumbers. As the name suggests, this vacuum sucks up both dry debris and water. Wet/dry vacs differ another according to their horsepower ratings and the volume, measured in gallons, that they can hold. You don't have to go all out on these (prices start at less than $50 for a 6-gallon model), and they can be used for all kinds of clean-up work. Sears' Craftsman line even offers a gutter-cleaning accessory kit for vacuuming or blowing out gutters without having to climb up a ladder.

Care and Feeding of Power Tools

Power tools are great time savers and are more fun to use than hand tools. Admittedly, that's a biased male viewpoint because we tend to like anything that makes a lot of noise while at the same time putting holes in things. These tools won't be fun for long if they're misused and abused, though. Avoid these practices:

➤ Lifting the tool by pulling on the power cord instead of the handle or body of the tool

➤ Applying too much pressure while in use, despite warning signs of the blade or drill bit slowing down, straining, or the motor giving off a burning smell

➤ Ignoring damaged cords

➤ Leaving the tool out in the rain

Tools don't ask for much. They're something like huskies and dog sleds. Treat huskies well and keep them fed, and they'll pull you and your sled until they drop. A power tool will keep going and going if you take care of it. I've run across homeowners who had 40-year-old electric drills, and they still ran like the day they came out of the box.

Pipe Dreams

Recharge your cordless tool's battery as soon as the motor starts to slow down. You don't want it to get completely drained before recharging, because this can damage the battery.

Taking care of your tools will also protect you at the same time. A frayed cord can lead to a short circuit, never a good idea when you're working. A dull blade or drill bit can cause the tool to slip and endanger your hands or legs by cutting into them instead of the wood you're aiming at.

Rent or Buy?

One of the main differences among power tools is the size of the motor. Professional, heavy-duty models have large motors and can reduce your drilling time through wood and masonry. Hand tools have their differences, too, usually in the quality of the metal components and the sharpness of the cutting edges.

If you have a lot of drilling to do through wood framing, you'll do the job a lot faster with a heavy-duty, $\frac{1}{2}$-inch drill, but these are expensive (usually around \$250 or more). A plumber will use this kind of tool regularly and can justify the expense of owning one. Renting one for one or two days makes more sense for a homeowner.

Before you rent anything, handle the tool and get a feel for it. A heavy tool can be uncomfortable for some people to handle for an extended length of time, so take this into consideration. You might be better off with a smaller drill that you can handle more safely. When I had nothing else available, I drilled through old, hardened floor joist with a $\frac{3}{8}$-inch drill without the sharpest of drill bits and still got the job done (not that I recommend this approach).

Plumbing Perils

Read over your rental contract carefully. Unless you take out optional insurance, you can be liable if the tool is damaged while in your possession. Most rental stores will check that the tool is working before it goes out the door. Be sure that you receive complete operating instructions if you're unfamiliar with a tool.

UL-Approved Parts for You

It would be unusual to run across an electrical component that isn't approved by the Underwriters Laboratories, but always check for its tag or stamp of approval on anything you buy, whether it's a flashlight, power tools, or an electrical device. A UL listing is your assurance that the product has been tested for safety. Receptacles, light switches, light fixtures, and appliances should all have UL approval.

Bear in mind that UL approval doesn't imply longevity or ease of installation. A cheaper, lower-end product will never be the equivalent of a more expensive product.

Tools You'll Probably Never Use

Plumbers have more tools than the ones we've listed, but if you ever need some of them, you'll be *way* beyond the do-it-yourselfer stage.

➤ Plumber's die and taps for threading replacement sections of steel or iron pipe

➤ Calking equipment for melting lead and applying it to the joints of cast-iron pipe used in the main stack or soil pipe

➤ A chain wrench for removing or installing a cast-iron soil pipe

Spare Parts

The following is a list of inexpensive spare plumbing parts:

➤ A variety of rubber and plastic washers for faucets and hoses

➤ Leadless solder for joining copper pipes

➤ Penetrating oil for frozen or rusted pipe threads

➤ Extra hacksaw blades

➤ Wax toilet bowl ring

➤ Old rubber inner tube for wrapping around minor pipe leaks

➤ Hose clamps to secure rubber patch around pipe leaks

➤ Pipe clamps (again for pipe leaks)

➤ Epoxy putty to seal leaks in pipe joints

➤ Flapper flush valve for toilets

➤ Teflon tape for sealing threads on steel pipe

Keep a flashlight in your toolbox as well. Also be sure to keep a supply of absorbent rags on hand and a drop cloth or sheet of plastic to place under the fixture being worked on.

The Least You Need to Know

➤ You don't need the costliest tools, but cheap tools are no bargain.

➤ Buy your tools according to the job at hand, not some vague notion of what you might need later.

➤ Handle all tools carefully, especially electrical tools.

➤ Buy the tools you'll use over and over again, but consider renting those for which you have only occasional use.

HEY

HELLO

HI THERE

Chapter 8

Pipes: Joining and Fitting

In This Chapter

➤ Pipe properties and characteristics

➤ Cutting and fitting pipe

➤ The advantages of copper

➤ You'll never work with galvanized or iron pipe

Plumbing is easy to understand, but most of us shy away from doing any plumbing work that we don't absolutely have to do, such as clearing the kitchen drain. Even then it's not uncommon to pour a bottle of chemical drain cleaner down and hope for the best rather than grabbing a plunger or disassembling the trap. This reluctance is understandable. If you do a sloppy job painting, you can always clean up and re-paint. Do a sloppy soldering job on your pipes, though, and you might spring a leak (and you might not be home when it happens).

Pipes are plumbing's roadways. You want them to be clear and planned out so that your water takes the most efficient route possible to get to your fixtures and faucets. Additions or repairs to your pipes can occur anywhere in the system. Some repairs to old galvanized pipes are often best left to a plumber, but others you can do yourself.

What's Your Type?

Some years ago during a construction project in Washington, D.C., workers dug up some of the original wooden pipes that ran under the streets. You won't find wood pipes in your house (you had better not, anyway), but you will find pipes made from one or more of the following materials:

➤ Plastic

➤ Copper

➤ Galvanized steel or iron

➤ Lead

➤ Brass

Plastic has been used for years as waste lines and drain pipe, and increasingly for water supply pipe. Copper is still the preferred pipe of choice for many plumbers, and galvanized steel or iron is no longer used in residential construction. Lead pipes were used decreasingly into the 1920s, and brass showed up as well but never met the economy or popularity of galvanized steel.

Despite their differences, plastic, copper, and galvanized pipe can be mixed and matched in your plumbing system through the use of adapters and various transition fittings. In an older house, you'll often find more than one kind of pipe. Of the three, plastic pipe is the easiest material to work with, especially for the do-it-yourselfer.

Pipe Dreams

Each pipe material—metal and plastic—has its own characteristics and properties that make it suitable for some purposes, but not for others. Each is cut, fitted, and joined according to these properties.

Check with your local building department before you install any plastic pipe. Additions previously done to your home using plastic might not be legal. Past work isn't necessarily a guide to what the code allows. Find out directly from the source, and you'll avoid problems later.

Working with Plastic Pipe

We discussed the different plastic pipes in Chapter 3, "Codes, Inspections, and Safety." Here's a quick recap:

➤ PVC is used for the drain-waste-vent system and is known for its light weight, ease of installation, and resistance to corrosion.

➤ CPVC is used for both hot and cold water supply pipe where permitted by code.

➤ PB is a controversial, flexible water supply pipe that is disallowed by a number of local plumbing codes.

➤ ABS is used as drain pipe and has also been the subject of lawsuits due to leakage.

In this section, we'll concentrate on PVC and CPVC pipe, although the cutting and joining techniques are similar for ABS pipe. Both PVC and CPVC pipe are rigid, easy to cut, and joined by a method called solvent welding. The application of the solvent liquefies each section pipe and fitting just slightly. When the sections are joined, the solvent evaporates and the plastic reforms, forming a tight bond between each section of pipe.

PVC drain traps and the pipe connected to them are not solvent-welded, however. They (the tailpiece, trap bend, and trap arm) are threaded and connected with slip nuts, so called because they can slip up and down a pipe until it's tightened.

Measuring Up

In many cases, you'll be installing pipe that's the same diameter, measuring across the inside, as your existing pipe. The length of the pipe will depend on your project, of course. Before you install any pipe, you'll have to cut it to size. To come up with the correct length, follow these steps:

➤ Measure from the outside edge of one fitting to the outside edge of the second fitting (called a face-to-face measurement).

➤ Measure each fitting for its socket depth (this is the length inside the fitting into which the pipe will be inserted).

➤ Add up the face-to-face measurement with the total from the two fittings.

The following figure shows the correct way to measure pipe for length.

Plumbing Perils

Measure twice, cut once. You can't redo a cut if you've cut a piece of pipe or wood too short. Take the time to measure again to confirm the correct length.

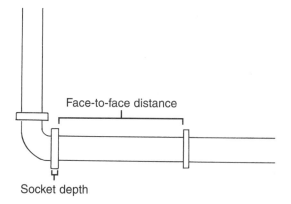

Face-to-face distance

Socket depth

Measuring up: figuring out the length of pipe you'll need for the job.

Cutting Plastic Pipe

Your cuts have to be clean, straight, and smooth. An electric miter saw, or "chop saw," will make these cuts perfectly. This is basically an electric circular saw attached to a stand, secured at one end, and capable of swinging around and up and down for angled cuts. Every finish carpenter owns one of these tools, which runs around $200.

The next best way to cut plastic pipe is to use a hand miter box and a fine-tooth saw. A tubing cutter can also be used, but saws are faster. One tool, a universal saw, is specifically made to cut plastic pipe as well as other materials requiring small, close-set saw teeth for smooth cutting results.

Pipe Dreams

Buy a fresh can of solvent rather than using one that has been sitting around your workbench for several months. Check the can for any hint of thickening or solidifying, and toss it if it isn't in a free-flowing liquid form. After all, solvent is cheap, but bad aren't.

What's That Thingamajig?

Dry fit just means to put pipes and fittings loosely together to check for size and length before you permanently connect them. The pipes are fitted together exactly as they will be installed and are marked with a felt marker or crayon for easy permanent installation. The direction of each fitting is marked with a line or arrow so that there's no guesswork while the solvent is drying.

After you've made your cuts, clean the cut end of the pipe with a knife blade by scraping the blade along the inside rim of the pipe or by rubbing it with fine-grit sandpaper until smooth. You don't want any errant bits of plastic interfering with the solvent weld. Run the sandpaper or a fine file along the outside edge, putting a slight bevel on it. Wipe the inside and outside of the pipe with a clean, dry rag.

Joining Plastic Pipe

Each type of plastic pipe requires its own solvent, so be sure that you're purchasing the correct material for your project. Here's how to join plastic pipe:

➤ After the pipe edge has been cleaned and the outside edge has been slightly beveled, dry fit and mark all the pipes and fittings together to be sure that they'll fit properly for your job.

➤ Wipe the ends clean of any grease or dirt.

➤ Apply primer to both the outside of the pipe and the inside of the fitting (this isn't paint primer, but primer specifically formulated for use on plastic pipe).

➤ Follow the drying instruction on the can (probably around 20 seconds), and then apply the solvent (PVC, CPVC, or ABS cement) to both sections, brushing on a healthy dose.

➤ Rejoin the pipe and fitting, twisting the pipe about a half turn to expel any air bubbles and distribute the solvent (be sure that you can see a line of solvent all around the connection).

➤ Check the solvent directions for drying times (they vary with the size of the pipe, but you must work quickly because it can set up in about a minute).

➤ Wipe off any drips of solvent that aren't benefiting the connection.

Hold the bonded section together for a couple of minutes. Don't run any water through the pipe until the bond has reached its full strength (check the solvent can for directions), which may take a couple days. If you find that you've made an error or have a leak and must undo a connection, consider simply cutting it out and installing new pipe with a coupling.

Applying solvent to plastic pipe.

Plastic and Metal Do Mix

As a rule, you can connect plastic pipe to metal, provided that the following points are true:

➤ You're inserting a *male plastic threaded fitting* to a *female metal threaded fitting*.

➤ No heat is involved in the connection (you cannot heat solder plastic).

On the other hand, you cannot connect a male metal threaded fitting or pipe to a female threaded plastic fitting. Plastic pipe can be connected to iron, steel, or copper pipe if the connection is done properly.

To join plastic pipe to copper, you need copper-to-PVC or CPVC transition fittings. Each section of the fitting is attached to its respective pipe first—the copper fitting is sweat-soldered, and the PVC fitting is solvent-soldered—and then the two sections of pipe are connected. You don't want to connect the plastic pipe to the copper connector and then solder it on to the copper pipe because the plastic will melt.

What's That Thingamajig?

Male thread is on the outside of a pipe or fitting. **Female thread** is on the inside. The male thread is inserted and screwed into the female. Think of it as the biology of plumbing.

Your situation might call for a different type of fitting than the ones shown in the illustration. Your sections of pipe might be set in place already, in which case a pipe union would be a logical choice. Or, you might be cutting into a copper pipe and installing a riser or vertical section of pipe and would need to install a standard tee first.

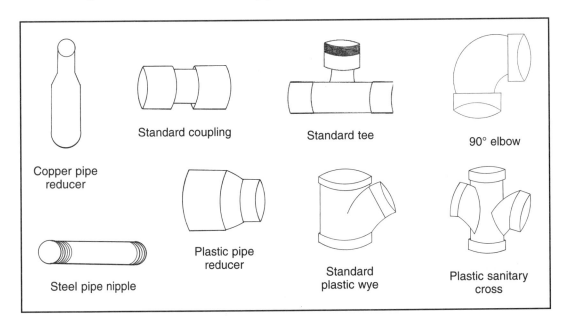

Typical plumbing fitting.

More Plastic: DWV System

Plastic is a great material choice for your DWV pipes. Unlike old galvanized steel or iron, plastic stays smooth on the inside and won't impede the movement of waste water. DWV pipe is laid out for as smooth a run as possible because waste water depends on gravity for its flow, unlike water supply pipes that are under pressure. DWV fittings are manufactured with less severe angles than fittings for water pipes to accommodate the flow of waste water. These fittings and pipes are solvent-welded, the same as plastic water pipes.

Flexible Pipe

The infamous polybutylene (PB) pipe (see Chapter 3) that was eventually banned by many local plumbing

What's That Thingamajig?

A **pipe union** is a connector that joins two sections of pipe that are fixed in place. The union consists of two or three threaded sections and includes both male and female thread.

codes is a flexible, easy-to-install alternative to rigid pipe. The main problem was not the plastic material itself, but rather the connections and fittings along with their installation process. If you have PB plumbing and are having problems with leaks, call a plumber and state specifically that it's PB pipe that needs to be repaired. Special tools are required for cutting the pipe, installing the fittings, and testing the connections.

Perplexed about PEX?

Despite being burned with flexible PB pipe, the plumbing industry continues in its search for alternative materials, as well it should. If the Wright brothers had given up after their first attempts at flight, we would never have known the wonders of frequent-flyer awards. Cross-linked polyethylene, or PEX, is the latest entrant in American pipe systems, even though it has been used in Europe for more than 20 years.

Cross-linked polyethylene results when links between polyethylene macro-molecules are arranged to create one large polyethylene molecule that is more resistant to temperature extremes and chemicals. PEX is created through one of several processes that start with polyethylene resin. One method adds chemicals to the mix, while another uses radiation. Not surprisingly, each manufacturer touts its product as well as its specific manufacturing process as being superior to those of its competitors.

PEX has a number of advantages, primarily its ease of installation. PEX has flex and requires far fewer fittings than rigid pipe. Each manufacturer offers its own system for fitting sections of pipe together (and each is the best, of course), and no soldering is required. Local codes might not approve of PEX, but various products have received approval by national code organizations such as BOCA and the IPC (see Chapter 3). PEX is available in both straight lengths and in rolls.

PEX has been successfully used in radiant floor heating systems and is increasingly being used for water supply pipes, depending on local code approval. Is PEX for you? There are two issues here, one structural and one emotional. Copper, for instance, has met the test of time, and its performance is predictable. We both know and feel that it's solid and

Plumbing Perils

One drawback to plastic drain pipes is that they're noisier than old steel and iron pipes, especially the main stack where water runs straight down to the sewer pipe. You can diminish some of this noise by packing fiberglass insulation around your main stack.

What's That Thingamajig?

The term **PEX,** for cross-linked polyethylene, is derived from the material and process used to create this new plastic. **PE** refers to the polyethylene itself, and the **X** refers to the cross-linking of the polyethylene across its molecular structure. The chains of the molecules are linked into a three-dimensional network that results in a hardy, durable plastic.

reliable. PEX and future plastic systems come with a certain stigma, a nagging question that asks whether they will last. After all, PB piping was code-approved and still didn't work out. In the end, you have to go with the material you're comfortable with, that your research deems appropriate, and that also fits your budget requirements.

It's easy to be reactionary and stick with only established materials, but I suspect that the future will be filled with plastic pipe in one form or another. Consider these questions: How did plumbers and builders react to the first copper pipe installations after years of galvanized steel and iron? Did they wonder whether these skimpy little soldered joints would ever hold up the way threaded fittings secured steel pipe? I suspect they did, but eventually copper became king and now is the standard against which all other materials are measured.

Why Plumbers Like Copper

Copper has many advantages as a pipe material and has been used in the plumbing industry for more than 60 years. Copper enthusiasts tout it over plastic, of course, and they have good reason for doing so. Some of copper's strong points include these:

➤ Copper meets or exceeds local plumbing codes in all 50 states.

➤ It's resistant to corrosive elements, high temperatures, and high pressure.

➤ It maintains its integrity as it ages.

➤ It's essentially maintenance-free during its service life.

➤ It's nonflammable in the event of a fire.

➤ Copper thaws easily in the event of a freeze.

And the list goes on: Copper is biostatic, so it inhibits bacterial growth; it can be tested immediately after installation; and it has low contraction and expansion rates associated with temperature changes. Finally, for all you ecology majors, copper is all natural and even has a premium value when recycled.

What's That Thingamajig?

The term **temper** refers to the hardness of metal. **Annealing** refers to the method by which the copper is softened.

Copper pipe for domestic plumbing is available in two categories:

1. Commodity tube or hard-tempered copper

2. *Annealed* or soft-*tempered* copper

Within the categories are three types of pipe:

K (heaviest)

L (standard)

M (lightest)

Types K and L are available in tube form or in rolls of soft-tempered pipe, but type M is available only in tube form. Copper tube is sold in set lengths. Annealed copper is soft copper that is sold in rolls and is regularly used for main water supply lines between a house and the water utility's main pipe. Soft copper in this installation allows for a single piece of pipe to be used and avoids the need to solder and connect various lengths of tubing for the same purpose.

Cutting and Joining Copper Pipe

The simplest way to cut copper pipe is with a tube cutter. These tools are sold in various sizes, so get the one that will take care of all your cutting needs. Be sure to buy a cutter that has an attached reamer for cleaning the burr or rough edge from the cut ends of pipes. If you have to work on installed pipes, you might need a mini-cutter for cutting in tight spaces. To cut your copper pipe with a tube cutter, follow these tips:

➤ Measure your pipe, including the sections that will be inside the fittings, and mark the surface of the pipe with a laundry pen.

➤ Be sure that the pipe is supported; if the heavy end bends down as you cut, you can get an uneven edge.

➤ Position the tube cutter so that its wheel is right on the mark.

➤ Tighten the cutter's handle just enough so that the wheel is pressed against the mark.

➤ Rotate the cutter around a couple times, and then tighten the handle some more.

➤ Continue tightening and turning until you've cut all the way through the pipe.

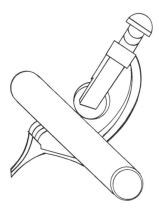

Cutting copper pipe with a tube cutter.

You need to tighten the handle only gradually. Use the reamer end of the cutter to clean the burr from the pipe by gently rotating it around a few times until the edge of the pipe is smooth. Check that your pipe is the correct length by holding it up to the area where it will be installed.

Joining Copper Pipe

Copper pipe is connected by soldering the joints with capillary fittings—that is, fittings that draw the heated, liquefied solder into the gap or joint between the fitting and the pipe. This type of soldering is referred to as sweat soldering. The end(s) of the pipe must be cleaned and prepared first. To solder copper pipe and fittings, you'll need the following materials:

➤ Steel wool or a fitting brush

➤ Flux and a small brush

➤ Leadless solder

➤ Propane torch

Pipe Dreams

If you're feeling uncertain about working with copper pipe for the first time, practice your cutting and soldering on some pieces of scrap pipe and some fittings. You can even make an art project out of it and build yourself a small sculpture. It's better to try out your new techniques with scraps instead of on the real job.

A fitting brush is a pipe-cleaning tool used on the inside of a pipe. You don't want any dirt or other contaminants to interfere with the solder and its adhesion. Steel wool, which is used on the outside of the pipe, can also be used to clean the inside. Be sure to turn the pipe downward and tap it to remove any bits of steel wool debris when you're finished cleaning it. Clean the inside of the fitting as well. Then clean the outside of the pipe until it's shiny, and wipe it and the fitting with a clean, dry rag.

Flux is used both to clean copper pipe and to prevent the oxidation of the joint. Solder is a soft metal alloy that, when melted, joins metal surfaces and seals the joint between them.

Time to Torch

It's easy to do sloppy soldering, but it's also easy to do a neat job. The following steps will get you through the soldering process and give you a tight connection with your copper pipes:

➤ After cleaning the cut pipe, apply a thin layer of flux to it.

➤ Install the fitting and wipe away any excess flux.

➤ Cut off a half-foot section of solder from the roll, and bend the end of it to a 90° angle.

➤ Light your torch, and hold the tip of the flame against the fitting, not the pipe. Move the flame all around the fitting as the flux sizzles, wiping away any excess flux (it takes only five seconds or so to heat one side of a fitting).

➤ Touch the end of the solder to the underside of the fitting joint; if it melts, it's ready to apply to the entire joint.

➤ Remove the torch and feed the solder along both sides of the fitting until it melts and flows into the joint, forming an even bead.

➤ Allow the joint to cool for a minute or so, and wipe off any excess solder with a damp rag (the joint will still be hot, and this heat will transfer to the rag, so be careful).

Soldering copper pipe.

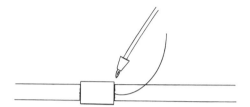

Wood and torches don't mix. If you're soldering a pipe that's already in place, be sure to protect any wall studs or floor joist by covering them with a flameproof metal sheet or even a piece of scrap drywall.

Check your soldering job: It should cover the entire joint without any gaps or small holes. If you see any, you can try to heat the fitting again and apply some additional solder. The ultimate test, of course, will occur after the water is turned on. Any leaks will have to be repaired, and this might require removing the fitting and starting all over again. To do so, you must reverse the process, but you need to add a few steps first.

Plumbing Perils

It's important to remove any stray flux from the finished soldered joint. Flux is corrosive and is useful only to prepare the pipe for soldering.

Unsoldering

When you have to take apart and resolder a fitting or remove an older, existing fitting from your copper pipes, the first thing you'll have to do it shut off the water and get it out of the pipes. Close the main shutoff to your water supply, and open all your faucets, especially those at the low points in the system. Take your propane torch to the fitting, and heat it until the solder melts and drips out. Grab the fitting with a channel-joint pliers—careful, the fitting will be very hot—and twist it free until you can remove it from the pipe. Be sure to clean out all the old flux and solder, and clean the ends of the pipe before reinstalling a new fitting.

Joining Without Solder

Soft copper—the stuff that comes in rolls—is joined with compression fittings or flared fittings. Compression fittings can easily be used on rigid copper pipe as well if you want to avoid soldering. Each fitting works by tightening against the copper pipe using some form of compression. The difference between the two is that a flared fitting requires a flaring tool to alter the pipe. You'll be more likely to run across a compression fitting in your home.

A compression fitting squeezes a metal ring or ferrule against the copper pipe by tightening a nut. A flared fitting forms a seal by tightening a nut outside the flared or expanded end of a pipe against the nose of the fitting. In each case, the compression of the pipe against the fitting forms a tight mechanical seal.

One advantage of these fittings is that you can install them while there's still a minor amount of water in the pipes, although you will have to shut off the water and drain the pipes first.

Galvanized and Iron Pipe

Most homeowners won't mess with old galvanized and iron pipe for good reason: Small problems can become bigger problems. Galvanized water pipe is threaded and connected with threaded fittings. The threads can corrode and rust over the years, requiring penetrating oil or heat from a propane torch to loosen the fittings. If you want to extend an existing galvanized system, it's simpler to run copper or plastic pipe off the galvanized pipe by means of appropriate fittings. The fittings are available at a plumbing supplier. Each one is threaded on one end and is tightened onto a galvanized pipe. The male threads are coated with a thin application of *pipe joint compound.* The other section of the fitting attaches to the copper or plastic pipe, respectively.

What's That Thingamajig?

Pipe joint compound is also called pipe dope. It's used to seal threaded fittings as they attach to steel or iron pipe, especially gas lines.

Cast-iron soil stacks and drain lines normally are connected at bell-shaped hub-and-spigot joints that were sealed with melted lead and a material called oakum that was forced into the joints. Cold caulking compound has substituted for the lead, although lead might still be used by some plumbers. This is another job best left to a plumber, even if all you want to do is tie into the stack by cutting a section out and adding a section of PVC. There are plenty of other plumbing jobs to keep you busy besides this one.

The Least You Need to Know

➤ If you have an old enough house, you might find galvanized, copper, and plastic pipe in use.

➤ Plastic pipe is the easiest to install, but it isn't allowed by all local plumbing codes.

➤ Copper pipe has proven itself in more than 60 years of residential use and is accepted in all 50 states for residential use.

➤ The key to successful soldering of copper pipe is to clean the pipe and fittings first and not to overheat them with your torch before applying the solder.

➤ Galvanized and iron pipe repairs and replacement can be tricky and often are best left to a plumber.

Chapter 9

Choosing Your Fixtures and Faucets

In This Chapter

➤ Sifting through the selections

➤ Know your fixture finishes

➤ Traditional materials vs. new materials

➤ Cleaning procedures

In plumbing, a fixture is a device that provides, holds, and disposes of water. Tubs, sinks, toilets, and lavatories are the fixtures we're most familiar with, but so are bidets, urinals, and whirlpool baths. The range of fixture choices and finishes is huge, with a price range to match. Your choices will include styles of fixtures, their colors, and the materials that compose them.

Traditional vitreous china and porcelain enamel-coated iron and steel have been replaced with plastics in some installations. In an odd twist, metal sinks, which were themselves replaced at the end of the nineteenth century by porcelain-enameled products, have made somewhat of a comeback. Stainless-steel sinks have long been used in kitchens and are now finding their way into bathrooms as well. Each type of material has its own maintenance and usage issues, and you need to be aware of them before you install a new fixture in your home.

There is also the question of installing additional fixtures beyond the basic necessities. Master bathrooms are getting second lavatories almost as standard equipment, and bidets and whirlpool baths are being added in new or remodeled bathrooms as well. Along with fixtures come faucets, from basic chrome to gold-plated models. Faucets all do the same job—some just do it more elegantly than others. This chapter will help you narrow down your fixture and faucet decisions.

Plumbing Perils

When looking at new fixtures at your plumbing showroom, look at the maintenance requirements as well. Some materials might require more cleaning or be more subject to wear and tear than you're willing to live with. Don't be fooled by a fancy design. Read all the accompanying literature before making your decision.

Pipe Dreams

There's no reason to stick with tradition in your bathroom if your enjoyment of your home is more important to you than the possible effect on resale value. You might take a very utilitarian view of your kids' bathroom and install an inexpensive stainless-steel kitchen sink and change it out after they're grown. On the other hand, you can redo a small powder room or half-bath to the max and install a copper sink and a marble vanity top. A little imagination can go a long way with your plumbing.

Basically, Fixtures Hold Water

Strip away the fancy designs, colors, and finishes, and all that a tub, toilet, sink, or lavatory does is hold water until we're finished with it; then it releases the water down a drain. We could even install a metal bucket with a drain hole in the bottom and call it a sink (this would certainly be cheaper than most sinks, but your plumbing inspector would be more than a little displeased with your selection).

Some materials, such as vitreous china, have stood the test of time. They have proven themselves to be the best, most affordable, and sanitary components and finishes available for plumbing fixtures. New ones, such as acrylics, have shown themselves to be worthy successors to some of the traditional materials as well.

In planning your plumbing changes and additions, fixture and faucet selection need careful consideration, if for no other reason than the expense.

It Can Get Ridiculous

Manufacturers and designers have pulled out all the stops to provide you with a bathroom that costs more than you might have once paid for a house. Some designers figure that you should spend as much as the equivalent of 10 percent of the value of your home on a new bathroom or bathroom remodel. It's always easy to figure such things when it's someone else's money paying the tab.

One result of the collaboration between designers and manufacturers is the various fixture collections, essentially matching suites of toilets, lavatories, bidets, and tubs in various styles. This is the no-brainer school of bathroom design and takes the guesswork out of choosing your fixtures. You might prefer to mix and match your fixtures, though, in which case you'll have no shortage of selections to choose among.

Bathroom Lavatories

Early sinks were made from metals such as lead, copper, and zinc. Thomas Twyford came along and developed the first one-piece ceramic lavatories in

England in the 1870s. The rest is history because this process became one of the standards for sanitary fixtures.

Bathroom lavatories are made from a variety of materials, including these:

➤ Vitreous china

➤ Enameled cast iron

➤ Stainless steel

➤ Cultured marble

➤ Polished brass

➤ Polished copper

Plumbing the Depths

According to Eljer Plumbingware Products, the company purchased an old dinnerware plant in Cameron, West Virginia, in 1907 and converted it to a sanitaryware plant. There the company developed the first vitreous china toilet tank. Eljer acquired more manufacturing plants and eventually consolidated vitreous china production in Ford City, Pennsylvania. Mansfield Plumbing Products now claims to have the largest, most technologically advanced vitreous china kiln in the world, with a hearth the length of a football field. It's located in Perrysville, Ohio. Kohler, on the other hand, claims to have the world's most modern vitreous china operation, located in Monterrey, Mexico.

Brass and copper are infrequent choices for lavatories and require a lot of maintenance unless you like the weathered, patina look. Lavatories come in different designs and shapes, including these:

➤ Vanity-mounted

➤ Wall-mounted

➤ Corner lavs

➤ Rectangle, oval, and round bowls

➤ Self-rimming lavs that sit on top of the vanity

➤ Undermount lavs that mount flush to the top of the vanity and are secured underneath

➤ Pedestal lavs

Pedestal lavs hide much of the plumbing behind a narrow pedestal, while vanity lavs conceal everything inside their cabinetry.

Toilet Choices

The one choice you don't have in a new toilet is the amount of water it requires per flush. The 1.6-gallon toilet is here to stay, but you have several models and manufacturers from which to choose. *Bidets* are also becoming more popular in new bathrooms, although some Japanese model toilets combine both functions of waste elimination and cleaning in one unit. (See TOTO toilets in Chapter 11, "Say Hello to Your Toilet.")

Toilet choices include these:

➤ One-piece units

➤ Two-piece units with a separate tank and bowl

➤ Standard gravity-assisted units

➤ Pressure-assisted units

What's That Thingamajig?

A **bidet** is a sit-down basin that's approximately the same size as a toilet, but is used for partial bathing. A spray of water washes the underside of the seated user. Bidets, which have long been popular in Europe, have their own water supply and drains.

One-piece models are costlier than their two-piece counterparts, but they're easier to clean and have a slicker design. Most toilets are made from vitreous china for easy maintenance and cleaning. Gravity-assisted toilets flush away wastes strictly by the force and weight of the water contained in the tank. A pressure-assisted toilet uses compressed air inside a separate tank to force the wastewater out and into the closet bend.

Blast from the Past

Some salvage companies and antique shops specialize in old plumbing fixtures, both refurbished and as-is. If only originals will do for you instead of reproductions, these suppliers will have items like these:

➤ Claw-foot tubs

➤ Galvanized tin tubs

➤ Pedestal sinks

➤ Unusual faucets

A lot of popular styles are still made as reproductions by mainstream fixture manufacturers. Be sure to look at these as well and compare prices with antiques. Compare also the condition of the older fixture with the new one. The cost of refurbishment could compel you to buy a new fixture instead.

Bubble Baths in the Claw-Foot Tub

Just about anyone who ever had an old claw-foot cast-iron tub longingly remembers sinking in for a long soak. Anyone who ever had to move one out of a second-story bathroom remembers it, too, because it seemed to weigh a ton. Old tubs were often made from enameled cast iron and, later, enameled steel because china was too fragile a medium for such a large fixture. During the manufacturing process, a *porcelain enamel* (a durable glass composition) is fused at a high temperature to the underlying metal tub. Enameled iron is very resistant to chipping, and enameled steel is less so. To chip porcelain enamel, as a general rule, you would have to dent the underlying base metal (it's safe to say that your tub won't be chipping anytime soon). Because of its weight, a cast-iron tub should really have a reinforced bathroom floor under it. And, in case your tub is ever subjected to juvenile terrorists, porcelain is both graffiti- and flame-proof.

What's That Thingamajig?

Porcelain enamel is a primarily vitreous or glassy inorganic coating bonded to a metal base by fusion at a temperature in excess of 800° Farenheit.

Whirlpools Don't Like Suds

One builder told me that he's had clients who were very keen on installing *whirlpool* bathtubs in their master bathrooms, but not so keen on them a year later. Clients told him the large tubs took a long time to fill and they didn't find that they offered all that unique of a bathing experience for the expense. You have to really like jetted tubs to go to the expense of installing and using one. A compromise version is available that's the size of a standard tub, but has the built-in water jets. Most whirlpool bathtubs are made from lightweight acrylic, and it's a good thing, given their size and the amount of water they hold. Stick with acrylic, and skip the fiberglass or marble tubs. Check the size and fit of the tub in the showroom (be sure that you and yours will be comfortable in it).

What's That Thingamajig?

The term **whirlpool** is often considered synonymous with **Jacuzzi**, but the latter is a trade name for a specific brand of jetted tub. The underwater jets in a whirlpool are considered to have therapeutic and relaxing effects.

From Tough to Indestructible

I recently read an article about a wood bath tub. It was constructed out of an extremely durable wood native to Africa, cost thousands of dollars, and was absolutely beautiful—as an art object. Art has its place and can even be found in the plumbing world in the design of fixtures and faucets, but the pesky world of sanitation dictates that these same fixtures and faucets must be easy to clean. Manufacturers offer a variety of finishes that are both eye catching and simple to maintain.

The most popular materials and finishes include vitreous china and porcelain enamel. Acrylics are fast becoming the first choice in bath tub and shower units, although you can choose fancier materials such as marble. An afternoon browsing in your local plumbing showroom will give you a better idea of the breadth of fixtures and faucets that will suit your plumbing needs. Keep maintenance and comfort in mind as you make your selections. A tiled shower stall might look really cool after you first install it, but not so when it comes time to clean the grout.

Plumbing the Depths

Even though vitreous whitewares are often referred to as porcelains, the ceramic industry would beg to differ. They maintain a distinction between true and technical porcelains and china, which is used for nontechnical purposes. China has a high glass content and good impact resistance. Technical porcelains are vitreous but also chemically inert. They are used in chemical ware and as insulators in electrical applications (sparkplugs are a good example). You'll probably never look at your china dishes in the same way again.

The Other China

The best traditional material for sinks and toilets is vitreous china, which is essentially impervious to water. Vitreous china is a fired and glazed clay that is worked to a highly polished, stain-resistant finish. This type of china comes under the product category of whiteware, which is broken down into three main categories by vitrification:

➤ Vitreous

➤ Semi-vitreous

➤ Porous

The greater the vitrification—the process by which the clay loses its porosity during the firing—the lower the material's porosity or capability to absorb water. Vitreous china's glassy surface readily holds water in containment, and that's why it's an ideal material for toilets and sinks.

It's a Plastic World

Traditional fixtures are like American cars up until the 1980s: heavy and with no plastic in them. Now cars are made with lots of plastic, and plumbing fixtures have followed suit with one major difference: No government mandates are forcing fixture manufacturers to produce plastic products. Cars had to lighten up to meet fuel conservation mandates. Plumbing fixtures simply offer another option with plastic fixtures. Let's face it, a vibrant economy embraces every new technology as long as it sells.

Acrylic tubs, showers, and whirlpools are lightweight, easy to install, and less expensive than traditional cast-iron models. The showers and tub/shower combinations also eliminate the need for a tub surround of any kind, such as tile. Acrylic is easily molded at the factory and retains water temperature well. Scratches in acrylic are repairable.

Fiberglass showers and tub/shower units are another alternative to traditional fixtures. These are not as durable as acrylic, however, and their finishes can fade faster than acrylic. Some proprietary products are added to the plastics mix as well. American Standard offers Enduran, an injection molding material engineered by GE Plastics specifically for kitchen work environments. This material resists scratching and staining, and its renewable surface can be cleaned with cleanser and scrub pads normally meant for pots and pans.

Cultured Stone

No, this isn't some intelligent geological formation. Cultured stone is made from ground-up pieces of real stone that are mixed with polyester resins. After the stone is shaped and formed, it's sealed with a protective gel coat. This is the particle board of the stone world. Corian by Dupont is probably the best known brand of cultured stone. It comes in any number of fixture variations, including lavatories, counters, and shower and tub walls.

Cultured stone is lightweight and easier to install than a natural stone such as marble. It will show scratches and is subject to chipping if you whack it with an iron skillet or similar object. Minor scratches and even burns can be sanded out with a fine-grade sanding paper.

Just Try to Break These

As the Penalware people say, "There is no better way to test for the function, strength, and durability of a plumbing fixture than the inside of a correctional facility." They should know—they manufacture stainless-steel, vandal-resistant fixtures for institutional use. Fixtures for correctional institutions must meet special criteria to withstand the less than socially benign activities of our ridiculously large prison population. In addition, Penalware comes with a Master-Trol Electronic Valve System, which can control up to 1,000 fixtures or more than 3,000 valves. Prison plumbing requirements are obviously more complicated than residential ones.

Kitchen Fixtures

Kitchen sinks used to be homely, but functional. However, new models are anything but plain. You can choose from single-, double, and triple-bowl sinks with sprayers, soap dispensers, built-in cutting and drain boards, and even colanders. Most people choose self-rimming sinks, but newer styles include above-the-counter models that sit on top of the counter rather than fitting into a hole in the counter and inside the cabinet below.

Traditional kitchen sinks were made from enameled cast iron and came in one color: white. More color options are available now, and you also can find sinks made from other materials, such as stone composite sinks. One of the biggest bargains is the stainless-steel sink. It's easy to take care of, doesn't chip, and doesn't show scratches as prominently as enameled sinks. Stainless steel is sold by its gauge; an 18- or 20-gauge sink will be of higher quality then a thinner-gauge steel.

A third type of sink is made from cultured stone and comes in a range of colors.

Plumbing the Depths

We once owned a house that was built in 1914 and came with a 1940s kitchen. The sink was enameled cast iron and had a high wall–mounted faucet. It was a shallow sink, but it was large enough to accommodate any size of pot or pan, including cookie sheets that could lie flat when cleaned. For all its fashionable shortcomings, it was a great sink and was much more practical than the one that replaced it. Style trumped practicality, but I wish we had combined the two and gotten a deep, wide, and long sink that wasn't broken up into separate bowls.

Cleaning Your Fixtures

Advances and choices in plumbing fixtures bring with them specific cleaning procedures. You can't pour out piles of scouring powder and start scrubbing away. (Well, you can, but you'll end up with scratches and worn finishes, in some cases.) All fixtures and faucets come with cleaning instructions when you buy them, but we'll summarize here.

Cleaning Materials and Methods

Surface	Procedure
Acrylic	Use a liquid cleaner or other nonabrasive cleaner. Spray or wipe this on the surface using plenty of hot water to scrub with a soft towel or rag. Rinse thoroughly when finished, and dry with a clean cloth.
Fiberglass	Same as acrylic.
Lacquered brass	Clean with a mild liquid cleaner, warm water, and a soft towel or rag; rinse and dry.
Unlacquered brass	Same as lacquered brass. To polish and help inhibit tarnishing, wipe once a week with a small amount of ammonia or brass polish.
Chrome	Clean with soapy water using a mild cleaner; dry and polish with a clean cloth. Chrome will also polish up with a small amount of vinegar weekly.
Cultured stone	Use a nonabrasive cleanser and wipe dry.
Vitreous china	Clean with nonabrasive cleansers or mild liquid cleaners; regular scouring powder can dull the finish.
Porcelain enamel	Use a nonabrasive cleaner, rinse well, and dry.
Mirrors	Clean with a commercial glass cleaner or a mix of equal parts vinegar and warm water. Dry with a clean, lint-free cloth or paper towels. Some scratches can be removed with toothpaste (nongel type) and polished with a clean cloth.
Gold plate	Clean with a mild soap solution and water, and wipe dry with a soft cloth. Do not use any cleaner that contains ammonia. Rinse fixtures thoroughly, and wipe dry.

Rust stain removal is a category by itself. Old fixtures, and sometimes not-so-old fixtures, can have rust or iron stains as a result of the local water supply or eroding pipes. I'd like to thank Anne Field, extension specialist, emeritus (with references from Purdue Extension bulletin Iron Control for the Home) and the Michigan State University Extension for their timely article on removing these stains.

Rust stains come from iron in the water supply, which can only be permanently removed through installation of an iron filter. Occasionally iron is dissolved from rusting water pipes or mains by corrosive water. Iron stains can be removed by a weak acid solution, usually oxalic acid, which is

Plumbing Perils

Never mix chlorine bleach or toilet bowl cleaners with other household cleaning products, especially any containing ammonia. Doing so can release dangerous gasses. Stick with using a single cleaning product per application.

highly toxic and must be handled with care. Never use chlorine bleach, as this sets the iron stain. Fresh iron stains on plumbing fixtures such as sinks, bathtubs, and chrome will generally yield to treatment with heavy-duty cleaning compounds containing large proportions of trisodium phosphate. Apply cleaner with damp cloth, pad, or sponge. Rub discolored surface until stain is removed. Rinse. Wipe dry and polish. Heavy, stubborn rust stains can often be removed by oxalic acid stain-remover compounds such as Zud or a tri-chloro-melanine compound such as Barkeepers Friend. Follow directions carefully. For rust stains already set on bathroom or kitchen plumbing fixtures, dissolve oxalic acid crystals in hot water and add enough whiting or talc to make a soft paste. Apply this poultice to stain, and let dry before removing. Rinse and polish. Use with care—it is poisonous.[1]

Plumbing the Depths

Mark Hetts, the columnist known as Mr. Handyperson, swears by Barkeepers Friend, a cleanser and polish for an assortment of surfaces, including stainless steel, porcelain, tile, chrome, and even glass. This product removes rust, lime, and other water stains from fixtures—and cleans drum cymbals, too, for all you interested tympanists. Barkeepers Friend is manufactured by Servass Laboratories (1-800-433-5818 or www.barkeepersfriend.com).

Faucet Facts

With faucets, the price often reflects the quality of the product. As a very rough measure, pick up two faucets for a weight comparison. The heavier of the two most likely will have more brass and stainless steel construction along with ceramic cartridges, all materials that are long lasting. Highly engineered faucets will have fewer moving parts to wear out and also will have exceptional finishes. You are also paying for a faucet's design; exotic-looking faucets will have a higher price tag.

[1]*This information is for educational purposes only. References to commercial products or trade names does not imply endorsement by MSU Extension or bias against those not mentioned. This information becomes public property upon publication and may be printed verbatim with credit to MSU Extension. Reprinting cannot be used to endorse or advertise a commercial product or company. For more information about this data base or its contents, contact cook@msue.msu.edu.*

You have a few basic choices when you pick out a faucet:

➤ Do you want one handle or two?

➤ What kind of finish do you prefer?

➤ How high do you want the spout?

➤ What kind of design or style do you want for a particular fixture?

Consider the spout height, even in the bathroom lav. A high spout in the kitchen means that it's easier to fill a tall pot or bucket. Mostly, though, faucet preference is a matter of design and taste. Look at plenty of samples before making a decision. The faucet must fit with the number of holes drilled in your sink.

Plumbing the Depths

Some new faucets come with a lacquered finish to protect the underlying metal and prevent tarnishing. Delta even offers a Brilliance line whose finishes are guaranteed forever while being resistant to most household cleaning products, including steel wool! Brass finishes that are not lacquered will require eventual polishing.

One safety feature you'll find in new tub and shower faucets is the pressure balance valve that is required in new construction. This valve adjusts for fluctuations in the water temperature by maintaining an even temperature within a small degree range so that you don't get shocked or burned. Fluctuations can occur when another user in the house turns on the water at a different fixture, drawing some away from the shower.

The Least You Need to Know

➤ There's no end to the choices of fixtures and faucet combinations available for your home.

➤ Be sure that you know the maintenance and cleaning requirements for your fixture's finish.

➤ Acrylic tubs and showers offer many advantages: price, ease of installation, and easy cleaning.

➤ The main differences between one quality faucet and another are the style and the design.

Part 3
Fundamental Fixes

Any time you combine running water with human beings, you have the potential for breakdowns. Well, not with the humans so much—at least, not in terms of their relationship with their plumbing, but with the plumbing itself. Even a rarely used faucet can eventually leak when a washer deteriorates after being under pressure for years and years. Pipes and faucets are always under pressure from water, and pressure brings breakdowns.

Drain and waste pipes aren't under pressure, but they are subject to clogs from all the gunk we put down them. Drain lines don't ask for much, just that we respect their limits and not see how much waste we can force down them at one time. Most clogs are easy to clear, and often the only tool you need is a plunger, a paragon of low-tech ingenuity.

A leaking pipe is not a signal to panic unless it's a burst main water supply pipe and you can't shut it off. Most leaks are minor but still need to be repaired before they move up in status to moderate or major. This part shows you how to replace a washer or the workings of a washerless faucet, and also walks you through some quick and easy pipe repairs. As a homeowner, you should be able to take care of the worst of your leaks and temporarily stop the biggest ones until you or your plumber can render a permanent repair. A little knowledge beats a flooded basement every time.

Chapter 10

Clearing
the Clogs

In This Chapter

➤ Clogs and their causes

➤ Plungers and augers, the uncloggers

➤ Clearing kitchen and lavatory drains

➤ Different tub and shower drains

➤ Avoiding clogs before they happen

Your DWV system is a simple design that works extremely well when you use it according to that design. A bathroom lav drain, for instance, is designed to receive soapy water, diluted toothpaste, shaving cream, and maybe an errant hair or two from time to time. It's not designed for wads of long hair, dental floss, or old bandages. Put those kinds of items down your bathroom drain, and you can expect a rebellion to ferment inside the trap or farther down the waste pipe.

Plumbers often are called in to clear out backed-up drains. A homeowner may have given up trying to clear the clog, while others simply call in a panic because they have no idea how to undo the mess themselves. Sometimes you *do* need to call a professional, but it should never get that far. Common sense (a fine parental term we all grew up with) and a little knowledge of your DWV setup will keep the pipes clear and the waste flowing.

This is a family affair, though. It doesn't do much good if you're the only one in your house who knows what to put down a drain and how to use a plunger. Everyone in your household should understand their dependent relationship on your plumbing and how to treat it. We'll discuss this along with unclogging procedures in this chapter.

Clogs Are Never Timely

It isn't true that sinks and toilets back up only during holiday parties, but it is true that there are more opportunities for problems to happen when your plumbing has more people using it. Not everyone will be as careful as you. In fact, one toilet supplier told me he believed the greater the distance in ownership between a plumbing system and its user, the more likely it is to be ill-treated (hotel owners and college maintenance staffs can attest to this), even if unintentionally so.

Plumbing Perils

It's a bad idea to wash plaster and joint compound down a drain. The materials can harden and really cause a clog. Scrape your tools off first, and wipe with damp paper towels that can be thrown away. Wash any remaining residue off with a garden hose.)

Each fixture has different sources of its clogs. Kitchen drains probably take the worst abuse—outside of toilets—with everything from celery leaves to shortening trying to make its way to the sewer line. Bathroom sinks and tubs get hair, soap, small toys, and bits of cotton. Even utility sinks take a hit when we wash out gardening tools and putty knives full of drywall joint compound. Plenty of water chasing this stuff down the drain often saves us from ourselves, but not always.

Prudent use of a drain will help keep your plunger tucked away in a corner of your bathroom. When you do need it, you'll find out why it's called a plumber's friend (and it's got nothing to do with the social lives of plumbers—at least, I don't think it does).

How Does Your Kitchen Sink Drain?

Considering all the food scraps, grease, and salad dressing that go down a kitchen drain, an occasional clog shouldn't come as any surprise. A good strainer over the drain hole will prevent most solid items from going down the drain, but you still can develop a sufficient accumulation of grease and very small bits of food to form a clog. If you have an old galvanized drain pipe, the problem is compounded by the internal disintegration of the pipe itself, making it more prone to stoppage.

Your first move with a clogged kitchen drain should be to grab your plunger and follow this procedure:

➤ In a double sink, stuff a rag or small towel into the drain hole in the bowl you're not plunging (if you just put the stopper in, it can dislodge when you plunge).

➤ If your dishwasher drain hose is attached to your disposer, it sometimes helps to place two small blocks of wood on either side of it and clamp them to seal off the hose.

➤ Scoop out all but 2 or 3 inches of standing water from the sink.

➤ Place the plunger over the drain hole, and push and pull it up and down 10 times or so without breaking its seal at the sink; you can ensure a better seal by applying a small amount of petroleum jelly to the rim of the plunger.

➤ After the last push, pull up on the plunger with a good, strong motion.

➤ If the drain has cleared, run some hot water down it for a few minutes to clean out any residual food.

Plumbing the Depths

Plumbers sell drain-cleaning products as well as their drain-cleaning services. RootX (www.rootx.com) is a foaming root control product that kills roots in pipelines. Bio-Clean (www.bio-clean.com)is a nontoxic drain cleaner that, according to the manufacturer, attacks organic wastes such as grease, hair, food, and paper. It does not create heat or fumes, but it changes waste particles into water, carbon dioxide, and mineral ash.

You could try chemical drain cleaner, but a plunger does such a nice job without the waiting. Besides, if you use a drain cleaner and it doesn't work, you have to scoop it out of the sink before you try the plunger because you don't want to be splashing it around your skin or eyes. If you feel a little queasy about using your toilet plunger, buy a separate plunger for your sinks.

Stubborn clogs might be beyond the help of a plunger and require an auger or snake.

Getting Serious with an Auger

A clog that resists a plunger might be farther down the drain pipe or really packed inside the trap. In either case, the trap will have to be removed. It's a good idea to examine old steel traps for rust and corrosion while they're off and replace them if they're deteriorated.

Pipe Dreams

You can sometimes avoid removing the trap if it has a clean-out plug on its bottom side. Simply unscrew and remove the plug, and insert the auger into the hole.

To remove a trap, follow these steps:

➤ Place a bucket or pan underneath the trap.

➤ Remove the slip nuts that secure the trap (for metal traps, use an adjustable wrench or channel-joint pliers) while holding onto the trap to steady it; water will drip out as you loosen the nuts.

➤ Pull the trap down (more water will gush out), and push the slip nuts and washers up onto the pipes.

➤ If the trap is packed with food waste, clean it out and rinse it; check the trap arm, the length of pipe coming out from the wall, for food waste as well.

➤ If the trap is clear, the clog is farther down the line, and you'll have to clean it out with the auger.

Anatomy of a trap.

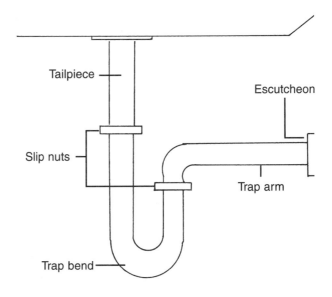

When a plunger isn't enough.

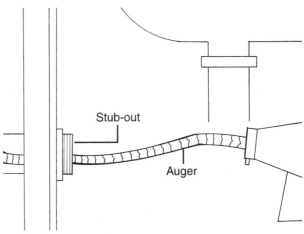

Augers work their way through soft food waste pretty easily. If you hit an obstruction, don't fight it, but rather back up the auger some and try again, twisting in both directions. With too much force, the auger can break through an old pipe or even a plastic one. When you're sure that the clog has been cleared, reassemble the trap by hand and tighten the slip nuts. If necessary, finish tightening with a wrench or channel-joint pliers until they're secure but not overly tight. Run hot water down the drain for several minutes to be sure that it drains freely and that the trap doesn't leak.

For clearing clogged disposers, see Chapter 17, "Kitchen Concerns."

Trapped in a Trap

Everyone knows someone who has lost a ring or other small item down a drain. If you notice soon enough and shut off the water, there's a good chance that it will be in the trap and will be easily retrievable. Remove the trap as described previously, and be sure to have a bucket or pan underneath to catch the water. If you inadvertently keep the water running after the item went down the drain, there's a better chance that it will be on its way to your local sewage treatment plant.

> **What's That Thingamajig?**
>
> An **auger** is usually thought of as wood-boring tool used by carpenters. It started out as the Old English *nafogar* (which related to a spear), moved up to the Middle English *a nauger*, and became incorrectly read as *an auger*.

> **Plumbing the Depths**
>
> There have been stories of rings being recovered by sewage treatment workers (they get caught in a screen) and being returned to their owners. If you're looking for those kinds of odds, you might as well buy a lotto ticket. These workers find all kinds of other debris of civilization in those screens, too, including sets of false teeth. These latter items aren't exactly claimed with any fervor by their recent owners, which isn't surprising.

Clearing a Lavatory Drain

A bathroom lavatory doesn't always get flushed as well as a kitchen sink, so hair and bits of soap can accumulate in the trap and beyond. If you notice that your sink is draining slowly, you might be able to head off a major clog by removing and cleaning the built-in pop-up stopper, which is a great gunk accumulator.

A pop-up stopper is opened and closed by a pivot rod underneath the sink. The pivot rod is attached to a clevis strap that is itself attached to the lift rod. Old lavs had a simpler deal because they often had a simple rubber stopper, sometimes attached to a small chain.

Stoppers are removed in different ways, depending on the sink:

➤ By simply lifting the stopper out of the drain

➤ By turning the stopper counterclockwise until it's free of the pivot rod

➤ By disconnecting the pivot rod from underneath the sink

Pivot rods disconnect by either loosening a retaining nut or, in newer versions, removing the plastic clips that secure the rod to the tailpiece of the drain. Once the pivot rod is disconnected from the stopper, pull it out and move it back and away from the tailpiece. Your stopper won't be a pretty sight, so be sure to clean it with hot water and soap. If it's full of hair, you've probably found your problem (and you should stop whatever you're doing that keeps dropping all that hair in the sink). Reassemble the stopper by reversing your steps, and check the flow of water down the drain.

Your drain exposed.

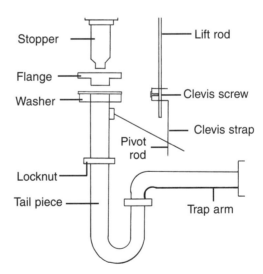

Taking the Plunge

The procedure for clearing the clog in a bathroom lav is similar to that for a kitchen sink. Instead of covering the drain hole in an adjoining bowl, you'll have to stuff a rag or towel into the overflow hole, which is usually opposite the faucet. Be sure that you have 2 inches or so of water in the sink to get a good seal with the plunger. Push and pull the plunger up and down 10 or 12 times; remove it and check the drain for flow. Plunging resolves a lot of drainage problems, but if you still have standing water in the lav, you'll have to remove the trap.

Trapectomy

Removing the trap of a bathroom lav is the same as with a kitchen sink. Place a bucket underneath to catch the water, and clean the trap after it has been removed. Many times the problem with a clogged drain can be found in the trap—removing and cleaning it should get your drain back to normal. If not, you'll need to use an auger or a plumber's snake to clear out the drain pipe.

Tub and Shower Drains

If you've ever found yourself standing in several inches of water while showering—not all that unusual in some vacation rentals, I've found—it's a safe bet that you have a clogged drain. The water might drain out eventually, but this is one time that slow and steady doesn't win the race. Another possibility is a poorly operating drain stopper that doesn't open far enough to allow water to quickly flow into the drain pipe.

Clearing a tub drain can be an involved procedure. Usually the trap isn't directly accessible (some old tubs have drum traps that are accessible from a cabinet behind the tub), and you must remove the stopper. Old tubs, like old lavatories, might have a simple rubber stopper that will just pop out. Others will have one of the following:

➤ A pop-up drain assembly

➤ A trip lever assembly with a drain strainer

➤ A metal stopper with a drain flange

The easiest of these three to deal with is the metal stopper. A metal stopper can lift and clear the drain hole, or drop down and seal it. The stopper screws into the flange; the flange screws into the tub itself. With the stopper removed, you can clear a tub drain with a plunger by the following method:

Pipe Dreams

Small manual augers are usually available at rental shops for a modest fee. These are more useful than the spring steel snakes sold at most hardware stores.

Pipe Dreams

Sometimes sticking a garden hose, with a nozzle on the end, placed inside a clogged tub drain can do the trick and loosen a clog. Pull the hose through a window, stick the nozzle down the overflow tube or drain, close the sink drain, and have someone turn the water on full force for a few seconds and then turn it off. Repeat this a few times if the water doesn't back up.

➤ Draw at least 2 inches of water into the tub to ensure a good seal with your plunger.

➤ Stuff a towel or rag into the overflow hole.

➤ Push and pull the plunger 10 times or so. (Rub a small amount of petroleum jelly on the rim of the plunger for a stronger seal between it and the surface of the tub.)

➤ Remove the plunger and test the drain; if it hasn't cleared, continue plunging and test again.

Some bathtub clogs won't respond to a plunger and will need to be cleaned out with an auger. Normally, you would put the head of the auger down the drain hole, but the flange to which the metal stopper is secured has a metal cross piece that cannot accommodate an auger head. The next best step is to remove the overflow plate and feed the auger right down the overflow tube and into the trap.

A metal stopper and overflow cover.

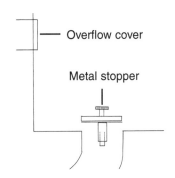

Overflow cover

Metal stopper

A clog can run, but it cannot hide from an auger!

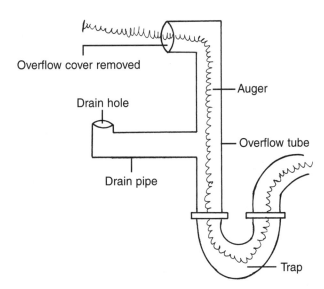

Overflow cover removed

Drain hole

Drain pipe

Auger

Overflow tube

Trap

132

Feed the auger into the overflow tube carefully, rotating it clockwise until you hit an obstruction. Back off slightly and try again. When all the obstructions are gone, slowly remove the auger and run some hot water into the drain.

Pop-Up Drains

A common tub drain is the pop-up style. Like the metal stopper, the pop-up drain also has a stopper right in the drain opening, but it's connected to a mechanical mechanism that raises and lowers it in and out of the drain hole. The trip lever, which controls the movement of the stopper, is part of the overflow cover. The entire mechanism should be removed, cleaned, inspected, and adjusted, if necessary, for proper draining and sealing.

To disassemble a pop-up drain, follow these steps:

➤ Move the trip lever to the open position, grab the stopper, and pull up, moving it back and forth a bit, if necessary. The stopper and attached rocker arm should all come out.

➤ Remove the screws securing the overflow cover and trip lever to the overflow tube, and pull out the cover and trip lever, along with the attached lift assembly.

➤ Clean off any hair and gunk from the drain pieces; if there's a lot of it, this might be the cause of your drainage problems and you won't have to clean out the drain pipe with an auger.

➤ Replace the O-ring on the stopper if it appears deteriorated.

➤ Run an auger down the drain pipe, if necessary; otherwise, reinstall the drain pieces.

➤ Run some hot water in the tub and test the drain; close the stopper and test for its seal.

➤ If the stopper doesn't completely seal the drain hole or doesn't open enough to drain quickly, remove the assembly again and adjust the position of the linkage accordingly (there will either be a set screw or adjusting nuts, depending on the model of drain).

➤ Assemble and test again.

Pipe Dreams

If you've disassembled a drain and found gunk in the stopper or other parts of the mechanism, go ahead and run your auger into the drain pipe anyway. Even if your leak appears to be caused by the drain components, you never know what might be farther down the pipe.

Bathtub drain assemblies attract lots of hair and bits of soap. A good way to keep them clear, as we've already mentioned, is to pour a kettle of boiling water down the tub drain once a week or so.

133

Pop-up drain.

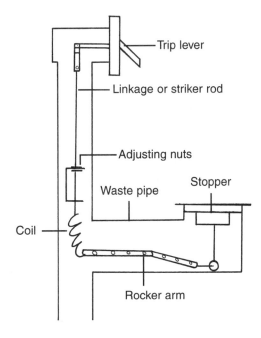

Trip-lever drain.

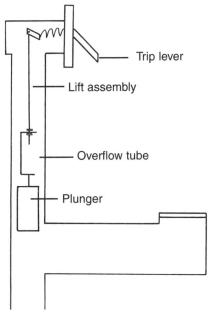

A strainer over the drain hole is a sure giveaway that you have a trip-lever drain. This system employs an internal plug or plunger inside the overflow tube that blocks a tube where it intersects with the waste outlet pipe coming from the drain. This assembly can be removed and cleaned by removing the screws that secure the overflow cover and trip lever, and carefully pulling out the linkage. Clean out any hair and gunk. If the assembly is fairly clean, clean out the drain pipe with an auger.

Passing the assembly back into the overflow hole can be a nuisance. Take your time, and ease it back in. Be sure to remove and clean the strainer as well.

Clearing Out the Main Line

Your main line is the big one that carries all your DWV system's waste to the sewer line (it's sometimes simply referred to as the sewer line). Harry Truman apparently had a sign on his desk that said "The Buck Stops Here," and your main line has a similar responsibility. Your blockage could be caused by household wastes or tree roots that have penetrated the pipe.

Main lines are designed to be cleaned out by the inclusion of a clean-out plug. The plug is often located at the bottom of the main stack or soil pipe. It also can be located farther along in the basement floor. You remove the plug with a pipe wrench attached to the square lug on top of the plug and then turned counterclockwise. Old cast-iron plugs can be tough to remove (after all, they aren't removed very often, and the threads can corrode). To remove a stuck clean-out plug from a cast-iron pipe, you have three options:

➤ Soak the threads with penetrating oil, and let it sit for half an hour

➤ Heat the threads with a propane torch

➤ Take a hammer and cold chisel, and hit the lug, trying to force it around counterclockwise

Plumbing the Depths

Tree roots have no manners when it comes to invading sewer pipes. These aren't necessarily the big roots that get in, but they're the small, fine roots that find their ways into pipe joints or cracks in clay sewer pipe. Once they're in, you'll probably have to have them cleaned out periodically because they'll come back.

Sometimes oil works, and sometimes it doesn't. Heat often works to free up reluctant metal threads. Heat them completely, and then run the torch around them a few more times. You may break the plug if you hit away with a hammer and chisel, but you can replace it with a new plastic plug. Wire-brush the threads on the connecting fitting first, and apply a light coat of petroleum jelly to the threads of the new plug.

Once the plug is off, the main line will most likely need to be cleaned out with a professional power auger, which can be rented, if you dare. I recommend hiring a professional drain-cleaning service instead—the experts have the machines and the various cutting heads to chop through just about anything, and they warranty their work.

Main line with clean-out plug.

A Few Defensive Measures

Once your drains and waste pipes are clear, some regular maintenance action should keep them that way. Some simple measures include these:

Plumbing Perils

It's okay to pour boiling water down a porcelain-coated steel or iron sink, but not down a china lav, a one-piece vanity top and sink combination, or a toilet. These can crack from the extreme temperature change.

➤ Pour a kettle or two of hot water down tub, shower, and kitchen sink drains once a week or so.

➤ Fill a sink with hot, soapy water and open the drain; the volume and weight of the water helps clear away bits of gunk in the trap and waste pipe.

➤ Pour some baking soda down the drain, and follow it with a couple cups of plain vinegar to cause a chemical reaction (you'll hear it fizzle); then rinse with hot water.

➤ Every few months, take a plunger to the drains you may have had trouble with in the past, simply as a precaution against any build-up.

Scalding water from a kettle will also help keep your fixtures clean.

The Least You Need to Know

➤ How you use your drains will determine when and if you get clogs or obstructions in your drain pipes.

➤ Plunging away the obstructions is often an effective way to clear your drains.

➤ An auger must be used carefully; some pipes can break if the tool is used too aggressively.

➤ Bathtubs have removable drain mechanisms that almost act as hair magnets; removing and cleaning the mechanisms will resolve a lot of tub drainage problems.

Chapter 11

Say Hello to Your Toilet

In This Chapter

➤ Understanding your toilet

➤ Common problems and their fixes

➤ Low-flow issues with tissue

➤ Replacing a toilet or tank

➤ Keeping it clean

The flush toilet is one of the single greatest sanitation forces ever created, but you probably don't see it that way when it's clogged and overflowing at two o'clock in the morning. Without toilets, we'd be reliving the lifestyle of pre-twentieth-century Americans with outhouses and water closets that emptied into backyard cesspools and privy tanks. So much for the nostalgia of a simpler, earlier way of living.

Toilets have a lot to do: They must hold water, hold us at times, flush away waste, and refill themselves with fresh water. A toilet's vitreous china bowl and tank will last for years, but the inner, working parts will need periodic replacement. Kids sometimes see toilets as a big toy, one that makes smaller toys disappear, at least long enough for them to cause a clog.

The working principles of a toilet are simple: When the handle is pushed, a flush valve lifts up and releases water from the tank into the bowl, forcing the water in the bowl down the drain (the closet bend). The tank and bowl then refill until the next flush. Any number of malfunctions, from small to major, can impede these actions. With a little toilet knowledge, you can ensure trouble-free operation for years.

This chapter covers toilet anatomy, repairs, and replacement. We'll even talk about cleaning and cleaners, which always come with the plumbing territory.

Working Principles

A toilet has to remove waste water and refill itself with clean water. All you really need is the toilet bowl itself, into which you could pour a bucket of water every time you needed to flush, but that would be inconvenient. The siphon action of the bowl sucks water out if and when a sufficient volume of incoming water with adequate force enters the bowl. In other words, if you repeatedly pour a cup or two of water into the bowl, nothing noticeable will happen. Pour a couple gallons in at once, though, and the volume of water and its falling action will empty the bowl until the siphoning action stops, allowing the bowl to refill. Pour the water in from a greater height, and the falling water will have even more force. Old, turn-of-the-century wall-mounted tanks, which were located 5 or 6 feet above the toilet bowl, performed their flushing function very well.

Buckets of water are okay to use during emergencies, but we prefer a standard flushing mechanism. When the tank handle is pushed, water is released from the tank into the bowl through a series of holes in the rim of the bowl and the siphon jet hole. The simple but elegant mechanics of a toilet assure us that we won't have to be using buckets every time we flush.

The Inside Story

Toilet design has stuck to the basics for decades. A vitreous china tank holds water, which flows into a bowl of the same material. The tanks and bowls are usually separate, although one-piece toilets are also available. The rim of the toilet bowl has a series of flush holes in it. Water from the tank flows through these holes when the toilet is flushed. It also flows through the siphon jet hole on the bottom of the bowl.

Unlike sinks and bathtubs, which have separate traps, toilet bowls have built-in traps to prevent sewer gases from entering the room. A drain hole at the bottom of the bowl allows wastewater to flow out and into the *closet bend* or waste line that carries the waste to the soil stack.

What's That Thingamajig?

A **closet bend** is the curved drain pipe located under a toilet. This is a wide, 4-inch pipe necessary to carry away toilet wastes.

Plumbing Perils

If you stand on a toilet, you can loosen it at the floor bolts. When reaching for something overhead and above the toilet, use a step-ladder or utility stool. The toilet could start leaking from a weakened connection with the floor, and you wouldn't even be aware of it.

The inner workings of a large number of toilets include the following parts:

➤ Float arm and float ball

➤ Ball cock or water intake valve

➤ Refill tube

➤ Trip lever

➤ Lift chain or wire

➤ Flush valve and flush valve seat

➤ Tank hold-down bolts

Pipe Dreams

Always check that the refill tube is actually refilling the tank through the overflow tube. Sometimes it gets dislodged after you've worked inside the tank. Also check that the handle isn't hitting the tank lid. This is an indication that the chain is too long and is allowing the handle to move too far up-ward. The chain will need to be shortened.

On the outside of the toilet is a tank handle that activates the flushing motion, a tank cover, and seat. A shutoff valve (you really want one of these with a toilet) feeds water into the tank through a water supply riser or supply line.

Each and every part of a toilet can be replaced, and some can be repaired by simple maintenance. Cracked tanks and bowls must be replaced, even if the cracks are small and do not appear to be leak-ing. They can eventually split open, and then you'll really know the meaning of leaking!

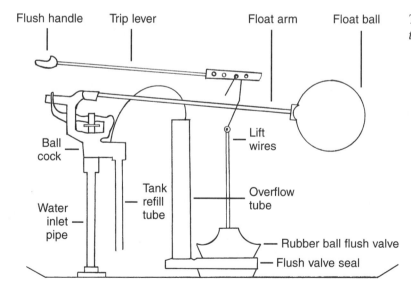

The inside story on your toilet.

Plumbing the Depths

Why does the water temperature change in some homes when a toilet is flushed and not others? It has to do with the way that water is distributed in the house. Old systems often had undersized (by today's standards) service lines and branch lines. If one fixture suddenly demands a lot of water, another fixture might suffer a noticeable deficit. A flushed toilet must refill its tank with cold water, which can be tough luck for anyone in the shower.

A lined toilet tank.

Mansfield Plumbing Products, Inc.

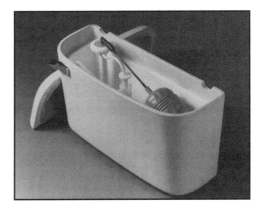

Flushing Action

A toilet flushes properly when all its working parts cooperate with each other. The following actions occur every time you flush a gravity toilet:

1. The tank handle is pushed, starting the flushing action.

2. A trip lever, connected to the tank handle, moves up, pulling a small lift chain or lift wire along with it.

3. The chain pulls up on the flush valve (either a stopper valve or a flapper valve, depending on the age of the toilet), allowing water to flow into the bowl.

4. When the water level drops to a predetermined level inside the tank, the flush valve falls back onto the flush valve seat, preventing any more water from going into the bowl.

5. The float ball at the end of the float arm also drops and activates the ball cock or refill valve, which fills the tank through a tank fill tube and the bowl through a bowl refill tube.

6. As new water comes into the tank, the float ball begins to rise again to the top of the tank.

7. When it reaches the top, the float arm closes the valve in the ball cock to prevent additional water from entering the tank.

8. An overflow tube directs any excess tank water into the bowl; the bowl refill tube sits inside the overflow tube and fills the toilet bowl independently of the tank.

9. The water level stops about a half- inch from the top of the overflow tube; no water should be dripping into the overflow tube, and no sounds should be coming from the toilet.

The series of events is different for toilets that are not gravity tank models because the inner works are different. Gravity tank toilets are still the most common, however.

Toilet Types

Toilets take clean water in and discharge wastewater. Three different types of toilets exist, and each performs these functions according to its own design. The three types are:

1. Gravity tank toilets
2. Pressurized tank toilets
3. Flush valve-operated toilets

The gravity tank toilet, with its bowl and tank, is the most common residential model. This toilet, which requires water pressure of only 10 to 15 pounds per square inch (psi), flushes away wastes when a volume of water is released by the tank. These are the least expensive toilets; one-piece models are more expensive than those with a separate bowl and tank.

A pressurized tank model stores water inside a tank. As the water enters, it compresses the air present in the tank. The compressed air releases pressurized water into the bowl and out the trapway. Pressurized toilets require water pressure of at least 25 psi to operate properly.

A flush valve-operated toilet is normally found in commercial settings with public restrooms. These toilets use a valve directly connected to the water supply instead of a tank. The valve controls the amount of water released over a set period of time during each flush. The advantage of this system is that it relies on pressure from a building's water supply and not from gravity. You'll probably never find these toilets in a private residence.

Pipe Dreams

Whenever you replace a toilet or do major repairs, install a shutoff valve if you don't already have one. You'll appreciate it anytime you have to do a future tank or bowl repair.

When Toilets Go Bad

A toilet is our most critical plumbing fixture. You can do without a shower and can always wash your hands with the garden hose, but when you have to go, your options are limited. An array of troubles can crop up, but most of them are homeowner-friendly and can be repaired with a few tools and a modicum of skills. A number of repairs call for shutting off the water and emptying the tank and/or bowl by flushing. You'll have to either shut off the water at the toilet itself or use the main shutoff to your house.

Every toilet troubleshooting guide looks like this one. If you've lived with the same toilets long enough, and especially if you have a family, some of these ailments will look familiar to you.

Toilet Troubles

Problem: The toilet does not flush.

Usual cause: The water has been shut off; the handle is disengaged or otherwise broken; the lift chain is disengaged or too loose; the refill valve or ball cock is malfunctioning.

Repair: Turn the water back on; replace or adjust the handle; hook up, replace, or shorten the lift chain; service the refill valve.

Problem: Water is dripping into the bowl.

Usual cause: The flush valve assembly has deteriorated.

Repair: Clean or replace the flush valve; replace the flush valve seat.

Problem: Water is flowing into the toilet bowl.

Usual cause: The float assembly needs adjustment or replacement; the handle needs adjustment; the water level is too high, and excess is flowing into the bowl through the overflow tube; the flush valve is leaking.

Repair: Adjust the lift wire, chain, float ball, or float arm as needed; clean or replace the flush valve.

Problem: The tank leaks.

Usual cause: The connection between the tank and the bowl is loose or deteriorated; the wall-mounted tank is loose at the flush elbow; leakage is occurring at the handle; there are cracks in the tank; the seal at the flush valve or supply line has deteriorated.

Repair: Tighten nuts and replace washers between the tank and the bowl; replace bolts, washers, and nuts; replace the packing of the flush elbow or the entire elbow itself; adjust the water level to below the handle; replace the tank; tighten connections or replace washers; replace the flush valve or its gasket.

Problem: The toilet bowl leaks.

Usual cause: The wax seal at the bottom of the bowl has deteriorated; the bowl has a crack.

Repair: Remove the bowl and replace the seal; replace the bowl and the seal.

Problem: Moisture or sweating appears on the outside of the tank.

Usual cause: Warm air is condensing on the outside of the tank.

Repair: Install tank insulation, or replace the tank with an insulated tank; ventilate the bathroom.

Problem: The bowl drains slowly.

Usual cause: Either the drain or the bowl is partially blocked; the water level is too low; the float ball needs adjusting; the flush holes under rim of the bowl are blocked.

Repair: Clear the clog with a plunger or closet auger; remove the toilet bowl to remove the clog; change the position of the float bowl to raise the water level; clean out the flush holes with a piece of thick wire or a wire hanger.

Problem: The bowl overflows.

Usual cause: Either the drain or the bowl is blocked.

Repair: Remove the blockage with a plunger or an auger; remove the bowl and clear the obstruction.

Problem: Water runs continuously in the tank (hissing sound).

Usual cause: Problems exist with the float; the water intake valve (ball cock) is damaged; the tank handle is sticking; the flush valve is leaking.

Repair: Adjust or replace the float assembly; repair or replace the water intake valve; repair the handle; repair the flush valve.

Note that siphoning can occur in a tank when an overly long refill tube is inserted too far into the overflow tube. A continually running refill tube is a constant waste of water. New refill tubes on water intake valves need to be cut to size so that they fit just over the top of the overflow tube and then attached.

Does Your Tank Need Anti-Perspirant?

If your toilet has this problem only when someone is showering or bathing, you need more effective ventilation in your bathroom; otherwise, you need to install an insulated liner (available at many hardware stores or home improvement centers), which is glued to the sides of the tank. Completely drain and dry the tank using rags followed by a hair dryer. Remove the float assembly, if necessary for access. Follow the kit instructions carefully, and allow the adhesive to dry overnight with the lid off the tank. Reinstall the float assembly and fill with the tank with water.

When Flush Valves Go Bad

Do the simple thing first, and jiggle the handle to reseat the flush valve. If your valve flapper is obstructed, flush the toilet with the tank lid off and see what, if anything, is in the way. Shorten the lift chain if it's too long and getting in the way by connecting it to a different hole on the trip lever.

If these solutions don't work, you'll have to check the flush valve. Turn off the water supply and flush the toilet. Wipe up any remaining water in the tank with rags or a wet/dry shop vac. Check the flush valve for deposits, especially if your area has hard water. Clean any deposits off the flush valve seat with a clean rag or the pot-scrubber side of your kitchen sponge. Examine the valve, and replace it if it's too deteriorated.

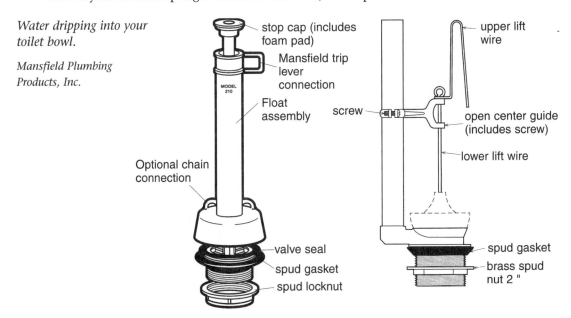

Water dripping into your toilet bowl.

Mansfield Plumbing Products, Inc.

To replace a flush valve:

➤ Shut off the water and flush the toilet, holding the handle down until you've drained off as much water as possible. Examine your flush valve.

➤ If you have a rubber ball (more accurately shaped like a toy top), carefully unscrew it from the lift wire. Screw on an identical replacement, which must be an accurate fit for a proper seal. If you have a flapper flush valve, unhook the lift chain from the trip lever (you'll be installing the new chain in the same hole).

➤ Remove the old flapper from the overflow tube. Some toilets require original equipment flappers. Your best bet is to take your old flapper, or ball, to a plumbing supply store along with the brand name of the toilet.

One variation on flush valves is the tilting flush valve, which consists of two plastic cylinders mounted at right angles to each other. An arm attached to the cylinders at one end has a disc on its other end that fits onto the valve seat. Replacement discs, both the snap-on and screw-attached versions, are available. Remove the entire valve from the tank (you either undo the bolt or use a screwdriver to spread fork blades at the hinge while lifting the valve), and clean and examine the flush valve seat. A deteriorated seat will need to be replaced.

Plumbing Perils

Be sure that the overflow pipe that comes with your new flush valve is the same height as the one you're replacing. If it's too tall, water will leak out of the handle hole if there's a problem with the water intake valve.

You have two choices to repair a flush valve seat:

1. Remove the tank and the overflow tube, and replace the seat at the bottom of the tube.
2. Use a Flusher Fixer Kit by Fluidmaster, Inc., which allows you to cement a new seat over the existing one.

Option 1: To remove the tank, shut off the water and empty the tank water. Disconnect the supply tube from the tank by loosening the coupling nuts with a channel-joint pliers or adjustable wrench. After it's loose, push the supply tube aside until it's just out of the way. Consider replacing the supply tube with a new braided stainless-steel line.

Next, loosen the tank hold-down bolts by inserting a screwdriver into the head of the bolt inside the tank to keep it steady while you loosen the nut under the bowl with an adjustable wrench. If the nut is difficult to remove, spray some penetrating lubricant on it and let that work its way into the threads for 5 or 10 minutes.

Removing the water supply tube.

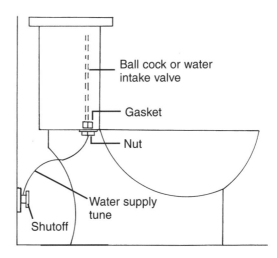

With the hold-down bolts removed, carefully lift the tank off and lay it on a towel on the floor, bottom side up. Remove the nut that secures the overflow tube with a spud wrench, and then remove the overflow tube. Either replace the entire tube or replace the seat only. Attach the overflow tube/flush valve seat assembly to the tank, reinstall the tank, and connect the water supply line and float assembly. Apply Teflon tape to the threads of the water supply line.

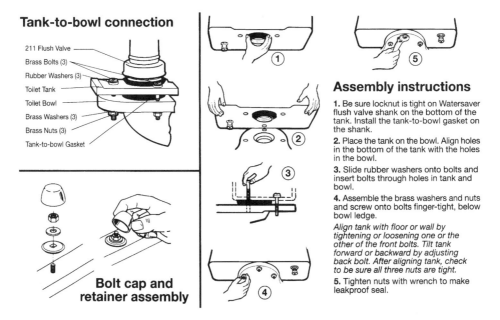

Tank-to-bowl connection

211 Flush Valve
Brass Bolts (3)
Rubber Washers (3)
Toilet Tank
Toilet Bowl
Brass Washers (3)
Brass Nuts (3)
Tank-to-bowl Gasket

Bolt cap and retainer assembly

Assembly instructions

1. Be sure locknut is tight on Watersaver flush valve shank on the bottom of the tank. Install the tank-to-bowl gasket on the shank.

2. Place the tank on the bowl. Align holes in the bottom of the tank with the holes in the bowl.

3. Slide rubber washers onto bolts and insert bolts through holes in tank and bowl.

4. Assemble the brass washers and nuts and screw onto bolts finger-tight, below bowl ledge.

Align tank with floor or wall by tightening or loosening one or the other of the front bolts. Tilt tank forward or backward by adjusting back bolt. After aligning tank, check to be sure all three nuts are tight.

5. Tighten nuts with wrench to make leakproof seal.

Installing a toilet tank

Mansfield Plumbing Products, Inc.

148

Turn on the water, sprinkle some food coloring inside the tank water, and check for leaks. All connections should be hand-tightened and then slowly tightened with a wrench. Excessive tightening will result in stripped threads and new leaks.

The water-intake assembly is an alternative to the standard ball cock unit and can be used to replace the latter. Instead of the standard float assembly, this mechanism uses a float cup that moves up and down the fill pipe. It's adjusted by loosening a screw on the side of the cup and moving the cup along the connecting rod that connects it to the ball cock arm. These assemblies are a cleaner design than a traditional ball cock.

Option 2: Your second option is Fluidmaster's Flusher Fixer Kit, which will repair many different types of flush valves (check the package for the exceptions). To use these kits, first turn off the water supply and empty the tank, wiping it dry. Remove the flush valve from the overflow tube. Clean the seat, using steel wool for brass seats and the pot-scrubber side of a sponge for plastic. Be sure that the seat is completely dry before installing the Flusher Fixer according to its instructions. Reinstall the flush valve, and attach the chain or lift wire.

Drip, Drip, Drip

These leaks can occur at the various connection points in the bottom of the tank, including at the tank hold-down bolts, at the ball cock or water intake valve, or where the water supply line comes in. Each one of these connections can come loose, and each has a rubber washer that can deteriorate. Place some food coloring in the tank, and examine each area for leakage to identify the source. If it's the supply line or hold-down bolts, try tightening them a quarter turn. You'll have to remove the tank to tighten the ball cock. You should replace the tank-to-bowl gasket or washer (also referred to as a spud washer) at the same time, in case it's the problem. Replace the tank hold-down washers as well.

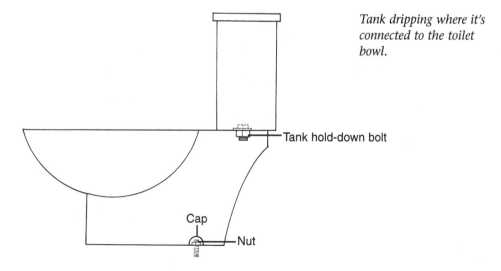

Tank dripping where it's connected to the toilet bowl.

Tank hold-down bolt

Cap

Nut

Whatever Floats Your Boat

If the water level is too high, the float ball is also rising too high and needs to be adjusted downward. Metal arms can be gently bent; plastic ones have an adjustment knob on the ball cock.

If the water level is low, adjust the float ball so that it will rise higher as the tank fills with water. The water should come within a half-inch of the top of the overflow tube. Using a small mirror, check the flush holes under the rim. Hard water deposits will restrict the flow of water into the bowl. Carefully clean these out with a piece of wire hanger or thick wire, taking care not to scratch the bowl. Clean out the jet siphon hole as well.

Adjusting a float ball.

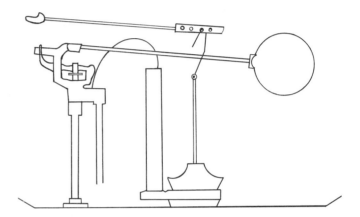

Slow Drain in the China

A plunger is called a plumber's friend for a good reason. For toilets, use one whose cup has a flange on the end (the flange can always be pushed up and inside for clearing sink drains) for a more powerful plunge. Several quick, short pushes followed by a solid push or two should clear most minor clogs. Flush the toilet, and repeat if necessary (be ready to turn the water shutoff in case of an overflow). If it's still clogged, you'll need a closet auger, which is a specially designed "snake" for toilets (don't use a regular drain snake because it can scratch the bowl). Insert the pointed end into the bowl, and turn the handle on the other end in a clockwise direction. Reverse direction when you hit an obstruction until the auger is out as far as it can go. Pull up and out to remove the auger, pushing back in and turning the handle as needed if it gets stuck. Use a plunger after the auger is removed to ensure that the blockage is gone. Flush the toilet to test.

In the worst cases, the bowl will have to be removed.

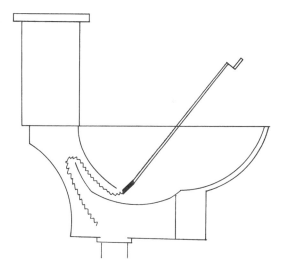

Slowly draining toilet bowl dilemmas.

Pulling the Tank and Bowl

Before removing your toilet, check that your local code allows a homeowner to do this job; it may require a professional plumber.

Shut off the water, spread some old towels around the base of the bowl, and empty excess water into a bucket using a throw-away plastic container or can, leaving the bowl about half full. If there isn't a shutoff or if it's frozen, remove the tank lid and lift the float up to shut off the water intake valve. Use a plunger or auger, as described previously.

If you cannot clear the obstruction after several attempts, the bowl will have to be removed. Flush the toilet to empty the tank, and then pour a bucket of clean water into the bowl to siphon off the remaining water (some will remain in the trap). You can use a wet/dry vac to suck any remaining water out of the tank and some from the bowl.

Disconnect the water supply line, remove the top of the tank, and store it out of the way. Remove the tank hold-down bolts and carefully pick up the tank, placing it on a towel or an old rug along with the tank top.

Remove the caps covering the bowl-mounting bolts, and place them with the tank cover. Undo the nuts securing the bowl to the mounting bolts, and apply penetrating lubricant if they're stiff and resistant. If the mounting bolts are too corroded, cut through them with a hacksaw. If the bowl has been caulked to the floor, break the caulk seal with

Plumbing Perils

Two people *can* pick up an entire toilet without disconnecting the tank if they're careful, but it's safer to remove the tank.

a putty knife, being careful not to cut into any vinyl floor covering. With the nuts removed, gently rock the bowl back and forth to break the seal at the floor, and then lift it up and away.

Either way, some water will remain in the toilet's trap; this can be drained into a flat pan. Place a rag in the drain hole to keep sewer gas from entering the room. Place the bowl on its side on an old rug or towel. The bowl sits on a flange (the closet flange) and is sealed to it with a wax seal or a plastic seal. Any old seal material has to be cleaned off the bowl and the flange with a putty knife and any corroded bolts must be removed (this job is best done with latex gloves, disposable or otherwise). If you replace the mounting bolts, note their exact location. You can install the new ones as you remove the old ones and tuck a small amount of the old wax ring around them to hold them in place until the bowl is reinstalled.

Removing a toilet tank and bowl. Overflowing toilet bowls might call for tough measures.

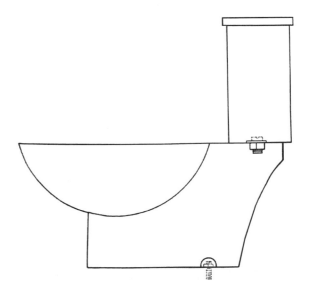

The best approach is to replace any fasteners, the water supply line, and the water supply line gasket as well. It's a small expense, and you'll be assured of a leak-proof fit. The floor should be examined for damage to the subflooring and the ceiling below. A longtime leak can even affect adjoining floor joist. A damaged floor will have to be repaired before the toilet can be reinstalled.

While the bowl is removed, clean the floor area underneath with alcohol or mineral spirits followed by warm, soapy water, and then rinse, and dry. Do the same for the bowl. While still wearing your gloves, probe the bend inside the bowl for the obstruction, which was the point of this whole exercise. If you can't reach it, insert the auger. When an obstruction can't be removed, the toilet will have to be replaced.

Place the new wax or plastic ring around the drain hole at the bottom of the bowl. Knead some plumber's putty into thin rope strips, and place them along the bottom rim of the bowl. The bowl or toilet now has to be lifted up and lowered over the hold-down bolts. This is a good job for two people, with one holding the bowl while the other guides it onto the bolts.

Once the bowl is back on the floor, press down on the seal, moving the bowl around slightly to help seat it properly. Coat the threads of the bolts with a small amount of petroleum jelly to ease future removal. Install the new nuts and washers until they're hand-tight. Finish tightening with a wrench with the usual caveat to be careful: The bowl can crack if the nuts are tightened too much. Clean off the excess plumber's putty and wipe the floor clean. Pour a few pails of water into the bowl to test for leaks. If any water seeps out, you'll have to do the job again. Seal the edge of the bowl against the floor with a thin bead of silicone caulking once you're sure that the bowl has been properly installed. The bowl should be snug against the floor and should not be able to be rocked back and forth. Then reinstall the tank and connect the water supply line.

Always install a new wax ring when you remove a toilet. You cannot guarantee the integrity of an existing ring, especially if you don't know how old it is. A new wax ring is a cheap investment.

Plumbing Perils

Don't keep flushing your toilet if it's clogged! The bowl will hold one tank load of water, but it won't hold a second one without overflowing onto the floor. Remove the clog, plunge the bowl a few times, and then flush again to confirm that the obstruction is gone. If you can't repair it right away, turn off the water at the shutoff valve.

Fill 'Er Up

Although a refill valve (a ball cock) can be repaired, it's generally simpler just to replace it with a new, up-to-code antisiphon ball cock. Your first step, after shutting off the water and draining the tank, is to remove the float assembly. Check the float ball. If it has any water in it, then it has a leak and must be replaced. A weighted ball won't float to the top of the tank and allow the water to shut off. The older, traditional ball cock was all metal, but most new ones will be plastic.

To remove the ball cock, shut off the water and empty the tank. Remove the float arm from the ball cock. Remove the water-supply line and move it aside. Hold the ball cock (you may have to tighten a channel locks to its base) and loosen the locknut on the underside of the tank with an adjustable wrench.

With the locknut removed, lift the ball cock out of the tank. If you install a metal ball cock, apply pipe compound to the threads before putting it through the hole at the bottom of the tank; plastic threads should be wrapped with Teflon tape or coated with Teflon paste. Secure the locknut by hand and then with a wrench until secure, but do not overtighten (the tank can crack).

Place the refill tube into the overflow tube, and install the float assembly. Reattach the water supply line and refill the tank. Flush it once, and readjust the float arm if necessary. If the water level is too high, the excess will keep running into the overflow tube or even out the handle.

Replacing a refill valve or ball cock.

Mansfield Plumbing Products, Inc.

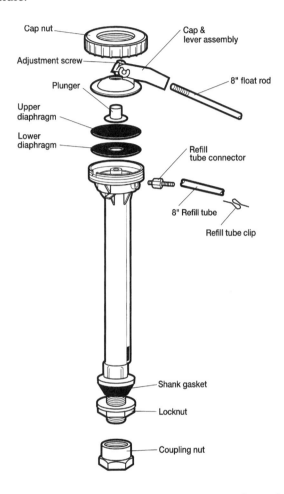

Areas with *hard water* might have more mineral buildup on valves, seals, and drain holes, which will require more attention and maintenance.

The Work Isn't Over Yet

Sometimes a simple plumbing repair involves more than the task at hand. A new toilet replacement, for instance, might require these tasks as well:

➤ Tile or floor repair

➤ Shutoff valve repair or replacement

➤ Repair of water supply line

➤ Modification of drain line pitch and/or venting system

➤ Disposal of the old toilet

A leaking toilet—especially one that has leaked undetected for a long time—can affect subflooring and floor joist, which might need replacement. Wet wood can rot and weaken and won't safely support a toilet. Some water supply lines are so old that they virtually can't be worked with confidently and should be replaced. You can see where this is going. Don't assume that you can simply pull a toilet out that was installed during Franklin Roosevelt's presidency and expect to pop in a new one without a hitch. You might not have any problems, but be forewarned that you might.

Low-Flow Blues

Way back when, before the rock-n-roll 1950s, toilets used lots of water to do their business, and they worked great. Seven-gallon flushes were common, and wastewater was carried away with great authority. Toilets used progressively less water as time went on, going to 5.5 gallons per flush (GPF) to 3.5 gallons and finally to the mandated (by the 1992 National Energy Policy Act) 1.6-gallon low-flow toilets we have today.

Early versions of low-flow toilets were anything but successful and often required two flushes to do their job (so much for water conservation). Manufacturers couldn't simply expect their existing toilet designs to work appropriately with less water. The humor columnist Dave Barry, who apparently has had an exceptional amount of experience with toilets, has written several columns on the subject, including one about an American couple who smuggled high-volume toilets from Canada to the United States for use in their home. So far, there is no word as to whether the border patrol will be training dogs to sniff out the errant china fixtures. Redesigned low-consumption

What's That Thingamajig?

Hard water refers to water that contains high levels of mineral salts such as calcium carbonate. These salts prevent soap from lathering properly and doing its job. These salts end up as deposits in your plumbing and can clog shower heads and toilet bowl holes and can build up on fixtures.

Plumbing Perils

In the worst-case scenario, your toilet might not be backing up, but your sewer could be. A first-floor toilet is often the closest opening to a sewer unless you have a floor drain in the basement. If the sewer is backing up, it will back up to its lowest opening. In this case, you'll need your sewer line cleaned out through the main clean-out plug.

gravity toilets often feature smaller diameter trapways, which require less water for siphoning, shallower depth, and less water surface area.

Pipe Dreams

If you're dissatisfied with your low-flow toilet, consider changing brands. Check out Terry Love's report on these toilets and their use in the real world at www.terrylove.com. A new toilet is less frustrating than regularly plunging out one that doesn't work.

Pipe Dreams

The easiest way to buy replacement parts is to take your old parts with you to your plumbing supplier. Jot down the name of the toilet manufacturer, too. You can find the name on the toilet bowl, the inside back wall of the tank, or the underside of the lid. Keep your old parts until the new ones have been installed and working.

It's illegal to install any toilet except a low-flow model in a residence, although that doesn't stop anyone from doing so. After all, the water police won't show up at your door measuring your toilet tank, but a building inspector will if your plumbing work requires an inspection. The new low-flow models have improved remarkably over earlier versions, although there are still complaints about them. The best of them, according to many plumbers, are the pressure-assisted models, which more than compensate for the lower water volume to remove wastewater from the bowl.

According to its manufacturer, the Sloan Flushmate Operating System is the most widely used pressure-assisted 1.6-gallon low-consumption system on the market today. The system's accumulator, a vessel inside the toilet tank, stores water under pressure. Air inside the accumulator, which compresses as the water supply line fills the accumulator with water, forces the water out in a vortex action. This effectively pushes waste through the trapway instead of depending on the traditional siphoning action. This system eliminates the traditional ball cock and flush valve. Instead of a tank handle, a button is pushed on top of the tank. One plus to pressurized systems is that they keep the toilet bowl cleaner through the force of the water during the flushing.

As advantageous as pressurized systems may be, they are louder than a standard gravity toilet. Small children might be startled the first time they use one, especially if they're used to quieter toilets.

Preventive Maintenance

The easiest repairs are those we can avoid or minimize. Regular maintenance, not exactly the mainstay of a society that prefers reacting to crises, will keep your toilet flowing (downhill, naturally) and calls to the plumber at a minimum. The following list should help keep your toilets in good working order.

Toilet Tips

Prevention beats losing use of your toilet every time. Common sense and a little up-keep will keep your toilet in uncommonly good shape.

Toilet Tips

Test your shutoff valve from time to time, turning it all the way off and then on again; if the valve is stiff, spray it with lubricant or apply penetrating oil.

Reconsider using toilet tank bowl cleaners; some of them damage rubber parts due to their chlorine content. Don't leave regular bowl cleaners sitting for too long in the bowl because they can possibly etch the porcelain.

Check the vent pipe coming out of the roof from time to time for obstructions from bird nests or other debris; a clogged vent can interfere with proper drainage.

A toilet is fragile; be careful when working around it with heavy tools.

Use a plastic toilet bowl brush; a brush with metal around the bristles can scratch the bowl.

A toilet is designed for bodily wastes only; use a waste basket for other paper products, bandages, sanitary items, and disposable diapers.

Don't store anything on the lid of the tank unless you're prepared to remove it from the toilet bowl if it falls in.

Don't pour hot water into the tank or bowl because they can crack from the temperature change.

Use only approved replacement parts when doing repairs (some must be replaced with original equipment manufacturer's parts only, while universal parts work in other circumstances).

Finding Those Pesky Leaks

Toilet leaks are often undetected, and as a result they are a primary cause of wasted water. Your toilet does give you some signals that it's leaking, including these:

➤ Sounds coming from the tank or bowl when the toilet is not in use

➤ The need for someone to jiggle the handle to stop the toilet from running, indicating a flush valve problem

➤ Water running over the top of the overflow tube, suggesting a leaking refill valve

➤ Water running down the sides of the bowl well after flushing

➤ Phantom flushing (the toilet turns the water on without anyone touching the handle)

157

Some of these leaks are harder to detect than others, but a few simple tests will let you know if they're occurring.

The Powder Test

A slow-moving leak in a tank or bowl is hard to see. Sprinkling some powder on top of the water in either section makes the leak more visible. A light sprinkling of talcum powder or cleanser will do the trick. A leak will break up the powder as it sits on the water.

A Test to Dye For

Some leaks are a bit too subtle for a powder test, so you need to color the water. You'll have to start with clean water, which means getting rid of any of the blue toilet cleaners inside the tank and flushing out the residue. You can also pour a bucket or two of water into the bowl to force out any blue water and replace it with the clear water from the bucket, retaining the water and cleaner inside the tank. If you're not using a bowl cleaner, remove the tank lid and sprinkle in some dark food coloring until it's well-mixed with all the water in the tank. Do not flush the toilet for at least half an hour. If any of the dyed tank water ends up inside the bowl, you've got a leak.

Now you know you've got a leak, but what's the source? The following steps will help you pinpoint the problem:

1. With the tank lid still removed, draw a line on the back of the tank with a waterproof laundry marker, just above the waterline.

2. Shut off the water to the toilet.

3. Leave the room and come back in 45 minutes.

4. Check the water level. If it has dropped, your leak is at the flush valve; if it hasn't dropped, it's at the refill valve.

Plumbing Perils

Don't put bricks in your older toilet to save water. Bricks disintegrate and can clog the flush holes in the rim or the siphon jet hole. Some toilets won't work that well if their water volume is decreased. If you must use something, use a capped plastic jug filled with water.

Cleaning Clues

Vitreous china is the material of choice for toilet bowls and tanks because it's a smooth, nonporous material that's easy to sanitize. Given the purpose of a toilet, you want to clean and sanitize it regularly. This is no place for a false dependence on in-tank bowl cleaners. Both the interior and the exterior of the tank need to be regularly cleaned with a disinfectant cleaner. Any cleaner claiming to be a true disinfectant will have an EPA registration number on the back label.

Spray cleaners are the most convenient, especially for toilet cleaning. Old standby cleanser works well for the inside of the bowl. Apply liquid disinfectant liberally, and allow it to sit for 10 minutes on all surfaces of the toilet, including the handle and base of the bowl. Wear a pair of dish-washing gloves or heavier rubber gloves, and work the bowl brush rigorously under the rim and the bottom of the bowl. Use a rag or a small scrub brush to clean the seat hinges, handle, and caps over the mounting bolts. Wipe the outside of the toilet dry, and flush away the water in the bowl. Be sure to wash the gloves and your own hands when you're done cleaning.

What's That Smell?

Toilets are designed to keep sewer gas out of the bathroom. If you smell gas, something's wrong. Possible causes include these:

➤ Low water level in the bowl, allowing gas to enter through the trap

➤ A deteriorated wax ring that no longer forms a complete seal at the base of the bowl

➤ A clogged toilet vent

➤ Clogged vents in other fixtures (which can allow gas to enter through tub and sink drains)

The water level can drop from leakage or evaporation in an unused toilet. A internal crack in the toilet bowl can allow water to seep directly into the drain. A clogged toilet vent can create a vacuum every time you flush, sucking too much water out of the bowl. Obstructed vents need to be cleaned out with a snake inserted into the vent at the roof, a good job for a professional.

What's Your Handle?

Toilet tank handles break one of the rules about tightening fasteners: The threads are frequently reversed, so you turn it *clockwise* to loosen and counterclockwise to tighten. If it's an older handle, try it clockwise first; if it isn't budging, try it counterclockwise. Apply some penetrating lubricant, if necessary, to loosen the nut that secures the handle.

Please Be Seated

Unless they're made out of stainless steel, toilet seats will not last as long as a porcelain tank or bowl. Aesthetics eventually demands their replacement. Hygiene mandates that wood ones either get replaced or refinished (pick replacement).

What's That Thingamajig?

For those readers who know only digital time pieces, **clockwise** means in the same direction as the hands of a clock. Fasteners tighten when they're turned clockwise and loosen when turned counterclockwise as you face the fastener. Keep this in mind if a fastener isn't immediately visible to you, such as one on the bottom of a toilet tank.

To remove the seat, flip up any lids covering the heads of the seat bolts. Older bolts won't have covers, but they might be corroded, in which case you'll have to soak them with lubricant and let them sit, possibly overnight. If you cannot loosen the bolts, they'll have to be cut off. Place a thin putty knife against the surface of the bowl, and carefully cut off the head of the bolt with a hacksaw. Once the old seat has been removed, thoroughly clean the top of the bowl with disinfectant, and then install the new seat.

Plumbing the Depths

Japan takes its toilets seriously. Researchers there are developing online toilets that will analyze urine samples and order a glucose or kidney test with your doctor, eventually recording day-to-day changes for your medical records. Toilet manufacturer TOTO offers a Washlet toilet, which will wash and dry your underside. Japan sponsors official Toilet Days, and several toilet Web sites originate there as well. Everyone needs a hobby, I suppose.

Septic Tanks

A septic tank is like a private sewage facility buried in your yard and it can be installed, when appropriate and allowed by local authority, in any home that isn't connected to a sanitary sewer line. A septic tank needs regular maintenance just like your plumbing system. If this maintenance is avoided ... well, let's just say that you'll become keenly aware of it in a very olfactory way.

A septic system processes raw sewage from a house sewer. It consists of three sections: a waterproof septic tank, the distribution box(es), and the drainfield or leaching field. Solid wastes stay in the tank as they separate from the liquids. Anaerobic bacteria inside the tank decompose the sewage, reducing it in volume to sludge that must be pumped out at regular intervals. These intervals are determined by usage, of course, which in turn is determined by these factors:

➤ The number of people using the system

➤ The amount and type of food waste going into the system

➤ The presence of a food disposer

➤ The capacity of the septic tank

Depending on your system and its variables, you may need to have it pumped out as often as every year or as little as every two to four years.

There is a science to setting up the sewer pipes and the tank so that the proper slope is obtained and that freezing is avoided inside the pipes, and this includes the location of inspection pipes and the pipe material itself. You want the work done by a competent, recommended contractor. Signs of septic system trouble include these:

➤ Toilets and drains backing up

➤ Drains smelling bad

➤ Wet or darker-colored grass above the tank, indicating a leak

If you have a septic system, find yourself a competent maintenance contractor to inspect and pump out the tank. This work cannot be done by a homeowner. Check with your local municipality to see if there are requirements for regular cleanings.

One Alternative

Flush toilets are not only convenient, but they're built around the American psyche. We want anything unpleasant to disappear while someone else—in this case, unseen sewage treatment workers—cleans up after us. Thanks to toilets, you'd never know that we're carbon-based food processors for whom what goes in must, in a different form, go out.

If you want to revolt against this artificial but easy-to-believe-in value system, you can always install a composting toilet (local plumbing codes permitting). A Web search will bring up all kinds of sites for manufacturers of these toilets that break down waste through the natural process of decomposition. Sewage and waste are primarily water by weight. Once the water in a composting toilet evaporates, only a small amount of compost is left over.

Manufacturers claim their products to be odor-free, all the while producing healthful compost for your garden. Organic material has to be added to the toilets to facilitate the creation of compost. A drum, in which the compost is created, must be turned

Plumbing Perils

Just because your septic system is independent of a municipal sewer system doesn't mean that you can treat it any differently when it comes to toxic chemicals. Keep all toxic and petroleum-based substances out of your septic tank. You don't want anything in the tank that will inhibit the bacterial disintegration of your wastes.

Pipe Dreams

Regular toilet flushing is a good idea, especially if you have hard water, because it helps prevent mineral buildup. Infrequent flushing to save water can backfire if you end up with clogged flush holes. If you're concerned about water conservation, get some recommendations from your local water department.

periodically to mix and aerate the material. One company builds its units out of materials tough enough to withstand freezing temperatures, even if the compost itself freezes, which is really a cheery thought.

A composting toilet makes sense for certain living situations, such as a remote cabin, but don't look for them to become a residential standard anytime soon.

When Natural Drainage Won't Do

Normally, for a fixture to drain, it has to be sufficiently higher than the sewer line so that gravity can do its job. What if you have a basement that is lower than the sewer line, but you want to convert it to living space and install a bathroom? You need to defy the laws of gravity, and you can do so with sewage ejector pumps.

Ejector systems typically consist of a basin, a pump, a float switch, a check valve, and a drain pipe connected to a sewer line. The bathroom fixtures drain into the basin; the float switch is triggered when the wastewater level reaches a predetermined height in the tank. The switch activates the pump, which forces the water up and through the check valve. The check valve allows waste water to flow out, but not back into the tank. The tank is vented through a vent outlet. You must have an electrical receptacle available to power the pump.

These systems work, albeit more noisily than a standard toilet flush, and are a workable option when you want to install a bathroom below the location of your sewer line.

The Least You Need to Know

➤ The basic design of a toilet, which is to hold water and flush away waste, has changed little over decades.

➤ A leaking toilet is one of a home's biggest water wasters.

➤ Each and every part of a toilet can be replaced or repaired if need be.

➤ A few simple tests will help you pinpoint the source of leaks within any toilet.

➤ Despite their shaky beginnings, low-flow toilets have greatly improved over the years.

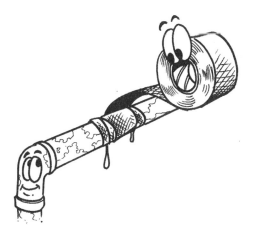

Easy Leak Repairs

We like our water to stay inside our pipes and fixtures until we're ready to let it go down the drain. Then we want it to go straightaway to the sewer and be out of our lives forever. When our plumbing system decides to circumvent this operation by leaking water from water supply pipes or drain and waste pipes, we respond. Leaks have to be patched up and repaired—and the sooner, the better.

Leak repairs are kind of situational. A once-a-minute drip in a basement bathroom that's used only by guests won't make it to the top of the "To Do" list. A burst main water supply pipe that's slowly filling your basement with water will get your attention much faster. Many leaks can be repaired temporarily. If the repairs take hold, they often become quasi-permanent repairs, although that's not the recommended approach.

Each and every pipe material requires its own repair procedures and materials to do the job right. You might not want to yank out and replace an errant section of pipe, but you can learn how to stop the worst of it, a good thing to know during an emergency. This chapter covers the different types of leaks you're likely to run across and tells how to repair them (at least until you can get hold of a plumber).

Leaks, the Hidden Story

A leak at the end of a spout or around a faucet handle often seems like more of a nuisance than anything we have to immediately repair. Individually, it doesn't amount to much, but collectively—when you add your leak to those of your neighbors— we're talking about major water usage and loss. Worse yet are undetected leaks inside your walls or under your house.

Leaks can be broken down into several categories:

➤ Highly visible leaks at spouts, faucets, and inside toilets

➤ Less visible leaks under sinks and in basement pipes

➤ Undetected leaks in crawlspaces, walls, and underground

➤ Minor leaks (minor drips)

➤ Large-volume leaks that require water to be shut off

Plumbing Perils

Small leaks typically become bigger leaks if left alone long enough. Undetected leaks can lead to all kinds of damage, even settling problems under your foundation. All leaks will run your water bill up without giving you any benefit whatsoever. Like any good parental directive, you should take care of leaks early on before they become unavoidable repairs.

Drain pipe leaks might not waste water, but they do introduce a less-than-sanitary situation around the area of the leak. Leaks in your sewer line can be very foul and must be repaired. Yes, that means digging— and hopefully not in the winter!

It All Adds Up

A leak means a water loss for you. A water loss is any water that comes into your plumbing system that never gets used, and that you still pay for. Aside from the waste of resources, it's a loss of money. A loss of one quart of water an hour amounts to 2,190 gallons a year. A more severe underground leak, one that wouldn't be noticed until the monthly bill comes, could lose a quart a minute, or more than 10,000 gallons a month! The exact loss depends on your water pressure and the nature of the break in the pipe (a loose fitting might lose less than a puncture, for instance). In any case, it behooves you to repair your leaks.

Leak Indicators

You might not see a leak, but you can see or hear evidence of one. The following signs can indicate a water leak:

➤ Higher than normal water bills

➤ Running toilets

➤ Musty odors and mildew

➤ Spongy plaster or drywall

➤ The sound of running water when all the water is turned off

➤ Moisture under carpets

➤ A wet yard near the house foundation, a hose bib, or in the area containing the main water supply pipe

A slight variation in your water bill can be explained by seasonal usage changes. A moderate to sizable change indicates a leak.

Plumbing the Depths

Even small leaks can be significant. In the case of a recent missile test, a plumbing leak held up the development of a $12.7 billion missile-defense system. The leak occurred in a small metal tube that carried nitrogen gas to refrigerate a pair of infrared sensor panels. When the sensors failed, the missile missed its target. At that point, the dominoes fell as further tests delays threatened the presidential decision to build the system at all. You never know where a plumbing leak is going to lead you.

Leaking by Design

The failures of polybutylene (PB) pipe fittings have been mentioned more than once in this book. As a result of these failures, various lawsuits have exacted big dollars from the pipe's manufacturers. Well, lawyers have probably exacted big dollars, but consumers are collecting a share of the settlements, too.

PB pipe is often gray and sometimes black in color. The problem fittings are gray or white acetal plastic insert fittings with aluminum or copper crimp rings. PB was used for both interior plumbing purposes and as main water supply pipes. If you have this pipe in your house and have leaks, you should call a plumber who has special tools for installing replacement crimp rings and equipment to perform pressure tests on the system.

Settlement funds have paid a percentage of actual damages caused by leaking PB pipe, as well as partial cost of replacement and repairs done to the system prior to

August 20, 1999. If you believe you're entitled to part of this settlement, contact the following firms:

Spencer Class Facility, 1-800-490-6997 or www.spencerclass.com (for interior pipe)

Brasscraft Claims Facility, 3600 Orchard Hill Place, Novi, MI 48376 (for interior pipe using metal insert fittings)

Consumer Plumbing Recovery Center, 1-800-356-3496 or www.kinsella.com/polybutylene (for exterior PB main water supply pipe installed between August 21, 1995 and August 21, 1997)

Plumbing Perils

In your kitchen, the dishwasher and ice maker are the more likely sources of leaks than the sink. Check on the dishwasher hoses running under the sink. Roll out your refrigerator periodically, and check the ice maker's copper tubing connection (you should vacuum back there once a year or so anyway, particularly the refrigerator coils).

When Copper Isn't King

In the wrong environment, even ever-durable copper won't hold up very well. Parts of California have such corrosive water that it isn't unusual to repipe a house whose copper system has deteriorated. Soil in some areas is also corrosive and has caused failure of underground pipes. Failed interior pipes have caused leakage severe enough to require repair of interior walls and exterior stucco.

Hard-to-Find Leaks

The toughest leaks to find are those that are in buried pipes, such as your main water supply pipe. Aside from simply losing water, water coming into your house could become contaminated by introducing dirt into your water supply at the point of the break. If you've discovered that you have a leak by checking your water meter, but cannot find it, you might have to check your underground pipes.

Plumbers do this work by using special acoustic equipment that detects the sound or vibration of pressurized water as it leaks from a pipe. This equipment varies in sophistication and usage and includes these items:

➤ Listening rods

➤ Sonoscopes

➤ Geophones or ground microphones

➤ Leak noise correlators

A ground microphone is used by plumbers to find leaking pipes under concrete foundations.

Plumbing the Depths

Mississippi State University decided to throw in the towel by closing its Blumenfeld Swimming Pool, 61 years after it was built, due to leaking underground pipes. The pipes, which were installed directly under the pool, were leaking more than 51,000 gallons of water a day. The university found that it would cost more to repair the existing pool than to replace it with a new one.

Critters Love Leaks

Undetected or ignored leaks can set up the kind of moist environment that says "Welcome" to various not-so-fun bugs such as cockroaches and carpenter ants. These are not benign invertebrate houseguests. They are another motivating force for you to repair your leaks, however.

Pipe, Fitting, and Fixture Sealants

Walk into a home improvement center and start looking at sealants, and you'll be there all weekend reading labels, ingredients, and instructions. The wonderful world of chemistry has brought us sealants to meet any specification. Plumbing requires different sealants for different materials and working situations (a pipe under pressure has different requirements than a drain line, for instance).

The following materials are used in plumbing repairs and new installations:

➤ **Teflon tape**   Easy to use and easy to misuse, Teflon tape seals metal pipe threads; the yellow version is used on gas pipe.

➤ **Pipe joint compound**   Known as pipe dope, this sealant lubricates and seals threaded metal pipe for water, steam, and natural gas applications; apply this liberally to the male threads.

➤ **Pipe joint compound with Teflon**   This material is intended for plastic pipe but can be used on metal pipes as well.

➤ **Plumber's putty**   This clay-like material seals under the drain and faucet where they meet a sink or tub. The putty stays soft and pliant.

➤ **Plumber's seal**   This two-part epoxy is sold by hardware stores and plumbing suppliers.

➤ **Silicone caulking**   Meant to seal sinks, vanity tops, and tubs at their perimeters, this sealant is not meant to seal pipes; caulking comes in tubes and requires a caulking gun to apply.

➤ **Wax ring**   This seals a toilet base to the toilet flange on the floor.

➤ **Solder**   This essentially creates a permanent seal between copper pipe and fittings, although it can be undone by heating with a torch.

➤ **Solvent**   This seals plastic pipe to plastic fittings.

The best sealants will fail to perform if they're inappropriately applied. Follow the directions on the package of each individual material, and don't skip any steps.

Plumbing the Depths

There seems to be some disagreement as to the suitability of Teflon tape in certain plumbing situations. Some plumbers use it as a sealant for any type of pipe material. My technical editor recommends using it on tapered steel pipe threads, but not straight-cut faucet threads. Some plumbers use it on plastic pipe, and others do not. Whenever it's used, it must be applied neatly without hanging out over the end of the pipe, where it can interfere with a faucet valve. To install, start wrapping the tape about an eighth of an inch from the end of the pipe, stretching it to pull it into the thread and overlap slightly for two or three layers.

Very Quick, Very Temporary Fixes

Any fix short of replacing a deteriorated pipe or fitting will be temporary, although this is a relative term. Some temporary fixes last for years (and years). A piece of rubber inner tube wrapped and secured around a slightly leaking drain pipe might stay put from the time your daughter first rides a bicycle until the day she starts driving.

The term *temporary* comes from the Latin *temporarius*, with *tempor* meaning *time*. It means that something is effective for a time only, and not an indefinite one. Think of these as short-term fixes until you can do a more permanent repair.

Such quick fixes include these:

➤ Homemade clamps and sections of rubber

➤ Pile clamps

➤ Epoxy repairs

Dads and granddads of a certain generation always had collections of odd bits of rubber, plastic, nuts, bolts, washers, and other miscellaneous stuff that few women would ever bother assembling. This might not be a very politically correct observation, but it is preponderantly true. If you ever buy an old house that was owned by someone who had one of these collections, you're likely to find a pipe or two repaired with a rubber scrap wrapped around a pipe and held tight with a wood clamp. This is a perfectly legitimate short-term repair that too often ends up being a long-term repair.

All leak repairs start with the same steps:

➤ Shutting off the water

➤ Draining off the water in the pipe by opening the nearest faucet

➤ Drying the area of the leak with a rag

Even if you don't have a personal collection of work bench odds and ends, you'll still be able to perform some temporary repairs to your pipes. There is a distinction between repairs done at joints and fittings and repairs done elsewhere on a length of pipe. Joints and fittings are a little more fussy about their repairs.

Plumbing Perils

Water supply pipes are under pressure. You want your repair to be one that will withstand and hold up under that pressure, especially at fittings. A simple pipe wrap repair can't be fully trusted, so be sure to perform a complete repair as soon as possible.

The Art of Homemade Fixes

All kinds of materials can be used to patch a hole in a pipe:

➤ An inner tube

➤ Clear plastic mailing tape

➤ An aluminum can

➤ Bailing wire

➤ An old garden hose

The standard do-it-yourself repair for a minor pipe leak calls for you to perform these steps:

➤ Shut off the water and drain the pipe.

➤ Wrap the pipe with a scrap piece of rubber.

➤ Place a small piece of wood on opposite sides of the rubber.

➤ Secure the wood, which will secure the rubber, with a small c-clamp that is ordinarily used in woodworking.

So what if do haven't got a c-clamp or rubber scraps? Use something else. Clear plastic mailing tape, for instance, is pretty tough stuff. Use it like this:

➤ Shut off the water and drain the pipe.

➤ Wrap a few layers of tape around the pipe, concentrating on the hole.

➤ Take a tin snips to an aluminum can, cut off the ends, and then cut again down its length until you have a flat piece of metal.

➤ Wrap the aluminum tightly around the tape.

➤ Loop some bailing wire around the can, leaving the ends of the wire about 6 inches or so in length.

➤ Twist the ends of the wire together with a pliers.

➤ Place a small, narrow piece of scrap wood inside this wire loop, and twist it around like a tourniquet.

➤ After the wire is tight against the can, tape the wood to the pipe by wrapping some electrical or duct tape around it and the pipe several times.

➤ Turn on the water and test.

Plumbing Perils

When repairing copper or plastic pipe, be careful how tightly you apply your clamps or other securing devices. Copper can bend and distort from too much pressure, and plastic can crack. You need your repair to be only tight enough to stop the leak temporarily.

This won't be particularly pretty, but it will stop the leak when you're in a pinch. Besides, what else are you going to use if it's Christmas Eve and your pipe springs a leak? There's a lot to be said for the ability to adapt the materials you have in the house to the problem at hand at still resolve it.

Hardware Helpers

You can do more refined repairs with a trip to the hardware store. Instead of bailing wire and plastic mailing tape, try a rubber patch held in place with two or more hose clamps. A hose clamp is simply a steel band that tightens as you turn its attached bolt. A cut-off section of an old garden hose split down its length also makes excellent patching material because it normally holds water under pressure.

To do a hose clamp repair, follow these steps:

➤ Shut off the water.

➤ Wrap the rubber patch or old hose around the hole (allow about 2 inches of material beyond the leak itself).

➤ Slip the clamps over the patching material, and tighten.

➤ Be sure that at least one clamp is over the leak itself.

➤ Turn on the water and test.

It's easy to do this repair and forget about it. Keep telling yourself that it's temporary, and mark on your calendar or personal digital assistant that you need to fix it soon.

Pipe Clamp Kit

This is a more official temporary repair. The kit comes with a rubber or neoprene patch and a clamp that fits all around the outside of the pipe and tightens with its own bolts. This is a pretty solid repair and will easily get you by until you can replace the pipe.

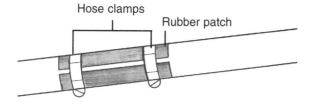

Hose clamps

Rubber patch

Pipe clamp repair.

Joint Repairs

Joint repairs can be as simple as tightening the fittings with a couple wrenches (on galvanized steel pipes). A slight turn can seal up a new leak. Too much of a turn, and you can get into real trouble with old water pipes: They can quickly break from the additional pressure.

If tightening a fitting doesn't do the job (it often does not), there's a "maybe repair" you can try. I call it that because maybe it will work, and maybe it won't. You can try filling the joint between the fitting and the joint with plumber's seal, a malleable epoxy that dries hard and is an exceptionally tough material.

To do an epoxy repair, follow these steps:

➤ Shut off the water and dry the fitting with a hair dryer.

➤ Mix the epoxy according to the package directions.

➤ Force the putty into the joint between the fitting and the pipe.

➤ Allow the epoxy to cure, turn on the water, and check for leaks.

On pressurized pipes, you might consider turning the water back on only partially to a lower pressure. There is no mechanical force such as a clamp holding the epoxy in place, so you should monitor it for leaks. If you have to be out of the house for a while and are uncertain about the repair, shut off the water until you return.

Fixing Waste and Drain Pipes

Old galvanized metal drain pipes can occasionally develop small, almost pin-hole leaks after years of service. If the hole is small enough, a really quick fix is to stuff a toothpick in the hole, break it off, and wrap several layers of duct tape or electrical tape around it. The wood will swell up as it gets wet and seal the hole. You should monitor the pipe for more leaks or weak points, and consider replacing at least the affected section at some point.

Metal drain pipes can also be repaired with the same clamps and epoxy used on water pipes. It's unusual for a drain line to simply start leaking, but that's not so for traps under sinks. Bump them one too many times when grabbing for something stored under the sink, and you can loosen the slip nuts that hold the pieces of the trap together.

Sometimes simply tightening the slip nuts will stop the leaking. Other times the washers that are present at each threaded section can become deteriorated and need replacement. If tightening the slip nut(s) doesn't do the job, you'll have to disassemble the trap and check the washers (see Chapter 10). Replace any that look the least bit damaged.

Plumbing the Depths

According to the Canadian Union of Public Employees (CUPE, www.cupe.ca), no fans of privatization of previously public utilities, "The worst example to date of the problems in water privatization occurred in the United Kingdom in 1995. During a devastating drought, NW Water imposed rotating water cutoffs at the same time as increasing water prices. Complaints were made that profits of $213 million were not reinvested into infrastructure, resulting in leaky pipes causing lack of water." CUPE claims that the water company was losing as much as 37 percent of its water supply, or 157 million gallons a day, through leaking mains and pipes.

Permanent Repairs

Some leaks will never be vanquished with a temporary repair, at least not for long. You will need to replace a section of pipe or fitting to affect the repair. Each type of pipe—galvanized steel, copper, and plastic—is repaired somewhat differently, but each starts with the removal of a piece of pipe or a fitting.

The most problematic to remove is old galvanized pipe. This pipe is exclusively connected by threaded fittings; the pipe and the fittings are installed in sequence, one section after another. The threaded end of the pipe turns clockwise into each fitting as it tightens. That means that the pipe is first inserted into one fitting and turned while the second fitting at the other end of the pipe is inserted onto the pipe, and it is then tightened. Removing a partial section of pipe due to a hole means that you will need a nipple and a union to repair the pipe.

Each section of pipe must be cut to size and threaded at each end. As the pipe ages, the fittings and pipe ends can corrode and can be difficult to disassemble. Even experienced plumbers must be careful working with galvanized pipe. New sections of pipe can be cut and threaded by your plumbing supplier and some hardware stores. This is a good job to contract out to a plumber who will bring tools to the job site for cutting and threading pipe. With some patience and care, you can work on your galvanized pipes, but with the caveat that they can be difficult.

Replacing Galvanized Pipe

Any cracked or leaking galvanized pipe will have to be loosened at each end to remove it and install a replacement. If the fittings are corroded, you can loosen them in one of two ways:

➤ Applying penetrating oil

➤ Heating the fitting with a propane torch

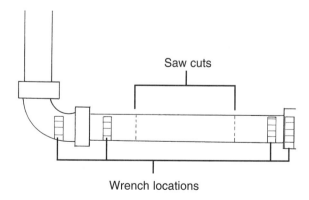

Corroded fittings should be replaced along with the pipe. To replace a section of galvanized pipe, follow these steps:

➤ Shut off the water and drain the pipe.

➤ Measure the existing pipe by noting the distance from the flange or outer edge of one fitting to that of the other; add another half-inch per fitting, or a total of 1 inch for the amount of pipe that will be inserted inside each fitting (your new pipe, nipples, and union must have a total length equal to these measurements).

Pipe Dreams

It's a lot faster to cut through galvanized pipe with a reciprocating saw and a fine-tooth blade. The drawback is doing it alone because the pipe can vibrate and possibly cause problems elsewhere in the system. You can avoid this by having a helper hold the pipe close to the fitting to steady it. A sharp blade will cut more smoothly than an old blade.

What's That Thingamajig?

A **nipple** is a short piece of pipe that has male threads on both ends.

➤ Purchase new pipe, nipple(s), and a three-piece union.

➤ Using a hack saw, cut the pipe about a foot or so from one of the fittings; if it's a long pipe, cut it 1 foot from the other fitting as well.

➤ Attach one pipe wrench to the first fitting and a second wrench to the pipe; grasping each wrench firmly, turn the pipe counterclockwise, being sure to keep enough pressure on the fitting so that it does not move.

➤ Remove the other section of pipe from the second fitting.

➤ Clean the threads of each fitting with a brass brush; if installing new fittings, clean the ends of the old pipe, apply pipe dope (pipe joint compound) to the threads, and install each fitting by twisting it on the pipe and tightening with a wrench while securing the pipe with a second wrench.

➤ Apply pipe dope to each threaded end of one nipple.

➤ Insert the *nipple* into one fitting, and tighten with a pipe wrench while holding onto the fitting with a second wrench.

➤ Separate the union into its three sections; slip the ring nut over the nipple.

➤ Apply pipe dope to the nipple threads, and screw on the hubbed nut, tightening it with your pipe wrench.

➤ Apply pipe dope to the threaded ends of the pipe (or nipple), and screw it into the other fitting.

➤ Screw the threaded nut section of the union onto this second section of pipe, tightening it with a pipe wrench.

➤ Align the two sections of pipe, and tighten the ring nut onto the threaded nut, using a pipe while holding the threaded nut immobile.

➤ Turn on the water and check for leaks.

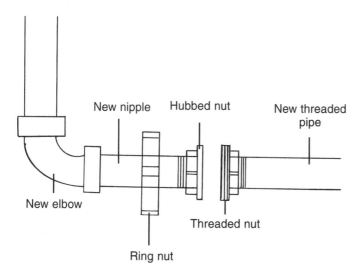

A small galvanized pipe repair that calls for many steps.

New nipple   Hubbed nut   New threaded pipe

New elbow

Threaded nut

Ring nut

In other areas of plumbing, too much tightening is a bad thing, but that's not so when dealing with galvanized pipe repair. Give the wrenches a good tight pull (you don't have to be ridiculous about it) and apply plenty of pipe dope to the threads.

Copper Capers

Copper pipes are soldered together at fittings unless a compression fitting is used that requires no soldering. A slip coupling is often employed to repair a section of broken copper pipe. Unlike galvanized pipe, you can easily slip new sections of pipe in just about anywhere. There are no concerns for threaded fittings because there aren't any. This is one clear advantage of soldering over threaded pipe.

The following steps will get you through a copper pipe repair:

➤ Shut off the water and drain the pipe.

➤ Cut out the section of damaged pipe with a tubing cutter or a hack saw (the tubing cutter will give you a cleaner cut); take out at least 6 inches, or 3 inches on each side of the hole.

Pipe Dreams

Even after you've shut off the water and opened the faucets to empty the pipe you're working on, some water might still remain. Place a bucket under the section of pipe or fitting you need to remove to catch any water still in the pipe. And stand back from the pipe while you're at it.

What's That Thingamajig?

A **slip coupling** is slightly larger then the pipes that it is connecting and can slide up and down the pipe freely. This allows it to facilitate repairs in the middle of a pipe run.

Slipping in a new piece of pipe with a pair of slip couplings.

➤ When you buy your replacement pipe, get some extra in case your first cut is incorrect and you end up with a piece that's too short; you'll need two slip couplings as well.

➤ Ream the cut ends of all the pipes to be connected, and clean the interiors with a pipe-cleaning brush.

➤ Clean the outside of all pipe ends and couplings with steel wool.

➤ Apply flux to the ends of the pipes and the inside of the slip couplings (be sure that the pipe is dry inside).

➤ Slide one slip coupling on each end of the old pipe.

➤ Place the new pipe between the two sections of old pipe, and move the slip couplings onto it until half of each coupling covers each section of pipe.

➤ Heat the couplings, and solder the connection.

➤ Turn on the water and inspect your work for leaks.

In lieu of soldering a slip joint, you have the option of installing a compression fitting. (See Chapter 8, "Pipes: Joining and Fitting.")

Leaking copper fittings are the result of poor soldering. To repair these, you'll need to shut off the water, drain the pipe, and heat the fitting until the solder runs out and you can remove it. After the fitting has been removed, the pipes and fitting need to be cleaned, fluxed, and soldered again (see Chapter 8).

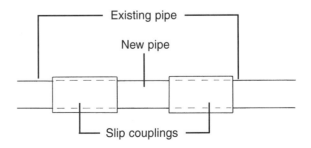

Plastic Patching

CPVC plastic pipe is among the easiest to repair: It's lightweight, inexpensive to purchase, and requires only solvent welding (see Chapter 8). It's repaired in a similar fashion as copper pipe. To repair plastic pipe, follow these steps:

➤ Shut off the water and empty the pipe by opening your faucets.

➤ Cut out the affected section of pipe (go about 3 inches beyond the hole) with a tubing cutter, a universal saw, or a hacksaw.

➤ Buy a new section of pipe and a pair of slip couplings.

➤ Cut the replacement pipe to size (approximately as long as the section you cut out), and clean the ends of all cut pile (see Chapter 8) with sandpaper and wipe clean.

➤ Brush the insides of the couplings and the outsides of the pipes with CPVC primer, following the product's directions.

➤ Apply CPVC solvent to the insides of the couplings and outsides of the pipes.

➤ Slide the couplings onto the old pipes and the new pipe into the gap between the pipes; slide one half of each coupling onto the new section of pipe, and hold the joined pipes together for a couple minutes while the solvent evaporates.

➤ Wipe away any excess solvent from the joint and the outside of the pipes.

➤ Following the directions on the can of solvent, turn on the water and check for leaks after an appropriate time has passed for the fitting to cure.

Be sure to dry-fit your new pipe and couplings before you apply the solvent. Check for a taught fit without any excessive gaps in the joints.

Small Leaks and Big Leaks

A leak is a leak is a leak, but not all are created equal. A small, visible leak at least announces itself and dares you to repair it. A gushing (relatively speaking) underground leak says, come and get me, once you can figure out where I am.

An underground leak should always be addressed. In the very worst cases, a damaged main can undermine a street or a house. If you don't want to tackle a leaking main water supply pipe, call a plumber. You can cut some of your cost by doing your own digging, but get the job done in any event. Same with small leaks. They're easy to ignore and just accept as part of the charm of an old house, but there's nothing charming about them. Remember, a leak is a sign of a malfunctioning plumbing system.

When Compression Valves Drip

A leaking compression valve, including shutoff valves under sinks, can often be repaired simply by tightening the compression nut or the packing nut on the valve stem. Tighten carefully! It doesn't take much, even as little at an eighth of a turn, to do the job. (See Chapter 13, "Start with a Faucet.")

Underground Leaks

The only alternative to digging up leaking underground lines is to shut off the water and not use them. That might work with a pipe that supplies a hose bib, but you can't do that with your main water supply pipe. Whether drain or supply pipes, you'll have to dig them up to repair them. Sometimes a new supply line can be installed without digging by attaching a new flexible copper line to one end of the old pipe while the pipe is pulled out of the ground by a winch or other motor. This can save a lot of work if your supply line runs under a driveway. Talk with your plumber about alternatives to digging.

Plumbing Perils

Frozen pipes can burst and cause a leak, so you want to thaw them out as soon as possible. If you absolutely must use a propane torch, place some nonflammable material against any wood surrounding the pipe. After all, if you start a fire, you won't have any water to extinguish it. And don't leave a space heater running unattended around frozen pipes, either.

When Leaks Get Really Bad

It's one thing when your kitchen sink spout drips, but what if your water utility has water losses? Utilities outside the United States lose water, and revenue, due to incompetent billing procedures, bad metering, and outright theft on top of normal leaks. All cities have some degree of leaking, especially if parts of the system are old. Managing water effectively and efficiently can allow a utility to put off the cost of increasing the capacity of the system. When water is better managed, consumption often goes down as does the amount of sewage.

Singapore, a city known for tough-love civic discipline, reduced its water losses due to leakage from 10.6 percent to 6 percent in six years. The entire water system is checked for leaks once a year, and meters are replaced every four to seven years. It's always inspiring when utility officials, who are quick to promote water conservation among its customers, adopt the same policies themselves.

The Least You Need to Know

➤ Every pipe, regardless of the material, is subject to leaking.

➤ Temporary repairs are just that: temporary. Plan on doing the permanent repairs as soon as possible.

➤ Old galvanized pipes are the toughest to repair and must be treated carefully to avoid further damage.

➤ Always test and inspect your repairs after completion; monitor temporary repairs until a final repair is done.

Chapter 13

Start with a Faucet

In This Chapter

➤ Identifying your type of faucet

➤ Compression vs. washerless faucets

➤ Simple washer repairs

➤ Different kinds of washerless faucets

➤ Spray attachments and aerators

You already know how to get water out of your sink, so now we'll discuss how to get it in. A faucet is kind of an end point valve—that is, instead of controlling the flow of water through your pipes, it allows the flow of water out of a pipe and through a spout. The spout section of a faucet is the water's end of the supply line.

Houses have different types of faucets. You won't find the same one on a utility sink that you would in a master bathroom (let's just say that it's unlikely). Each type of faucet has its own repair and installation techniques, but in the end all faucets do the same function and must be properly installed and maintained.

Before you can repair a faucet, you have to know how it's built and what replacement parts and tools are required for the repair. Most leaks start small, and that's when they should be repaired. If you let them go long enough, you'll end up shutting off the water at the shutoff valves (something just about everyone who's ever lived in a really old apartment house ends up doing at one point or another). Not only will you take care of the problem before it becomes a real nuisance, but you'll be saving water as well. Don't worry, faucet repairs aren't that complicated.

A Faucet for Every Purpose

Faucets come in many different styles, some of them almost whimsical in their artistry. Underneath the exterior appearance, though, are the mechanical workings. The faucets in your home are one of two types:

1. Compression
2. Washerless

A compression faucet does just what its name suggests: A washer is compressed to prevent water from flowing out the spout and then is decompressed to allow water to flow. A washerless (sometimes called a port-type) faucet uses different mechanisms to control the flow of water without requiring compression. A washerless faucet can be either a single-handled mixing faucet (these are always washerless) or a two-handled model. Compression faucets also come as two-handled or single-handled (your main shutoff valve is a compression faucet).

Plumbing Perils

When disassembling a faucet, lay the pieces on top of a paper towel or a rag away from the sink. Put them in order as they're are removed so that you won't lose track when reassembling. Also, plug the drain so that you don't lose any parts.

Washerless faucets, which have been around since the 1960s, are considered to be the longer lasting of the two types and incur fewer leaks, but there are still plenty of compression faucets around. The oldest faucets are always compression faucets. After about 15 years, either type can be ready for replacement.

Compression faucets work by relieving the pressure against a rubber washer, situated at the end of the stem, each time you turn the handle to the open position, thus allowing water to flow. Turn the handle in the other direction, and pressure on the washer prevents the flow of water.

Compression Faucet Repairs

Bathrooms are the usual location of two-handle compression faucets. The faucet parts—stem, washer, and valve seat, in particular—are under a lot of pressure every time the handle is tightened. It's not surprising, especially with old plumbing, that a drip sometimes develops in the spout. Drips can be the result of two problems:

1. A faulty washer
2. A damaged valve seat

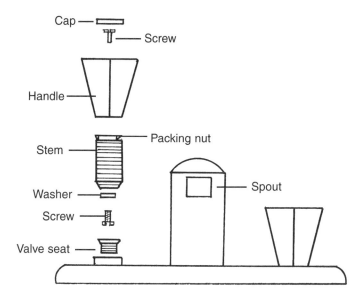

Compression faucet.

Cap

Screw

Handle

Stem

Packing nut

Washer

Screw

Valve seat

Spout

The first culprit, and the easiest to repair, is a faulty or deteriorated washer. To replace a washer in a compression faucet, do the following:

➤ Put your hand under the dripping water to determine whether it's the hot or cold water side that's the problem (another way is to turn off either the hot or the cold water and watch for drips; if the leak stops, the shutoff valve you turned has the problem).

➤ Shut off the water to that side of the faucet; turn the faucet handle and drain out any water.

➤ If your faucet handle has a plastic trim cap over it (marked C or H), remove it by gently prying with a small screwdriver or knife blade, or unscrew the cap itself if it's threaded.

➤ Remove the handle screw, and pry or pull the handle up and off the stem.

➤ Depending on the age of your faucet, you'll remove either a locking nut or a packing nut, either of which secures the stem to the faucet body; remove the nut with an adjustable wrench (if you have a large nut and a small nut, just loosen and remove the large nut).

Pipe Dreams

Even if the washer isn't the source of your leak and you have to do another repair, replace the washer anyway if it looks old. There's no point in not doing it if you've got the faucet disassembled.

New-style compression faucet.

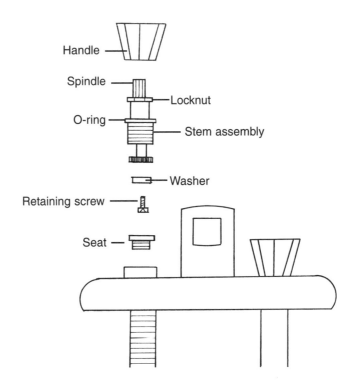

Handle

Spindle

Locknut

O-ring

Stem assembly

Washer

Retaining screw

Seat

➤ Pull the stem up and out, and examine the washer; if the stem is stuck, place the handle onto the stem (you don't need to screw the handle on) and give it a twist to loosen.

➤ If the washer is damaged, remove the brass screw that secures it to the stem. (If it won't budge, apply some penetrating lubricant and let it sit, even overnight if necessary.) Install an exact replacement washer and a new brass screw unless the existing one is salvageable. (If your washer doesn't seem to fit properly, take the stem to your plumbing supplier for a better match.)

➤ While the stem is out, clean off any corrosion or salts with very fine steel wool, being sure to wipe it clean before reinstalling.

➤ Install the new washer, apply a very light coat of heatproof grease to the stem, and reassemble the faucet.

If your faucet is still leaking, you should examine and repair your valve seat.

Meet Your Valve Seat

If your faucet still drips after the washer has been replaced, you probably have a damaged valve seat. This is the brass ring that is threaded on the bottom and screwed into the faucet body. If the valve seat deteriorates, the washer can't completely seal against the seat and prevent water from seeping through.

Plumbing the Depths

Faucets are pretty hardy pieces of hardware, considering the job they must do. It's not surprising that a leak develops on occasion. It isn't unreasonable to think about replacing faucets after they're decades old. Think of it this way: A $100 faucet that gets replaced every 20 years, whether it needs to be or not, costs almost a whopping 1.4 cents a day, minus any new washers or minor repairs. At these prices, faucets are terrific bargains.

Some valve seats are removable, and some are not and must be repaired in place. You remove a valve seat with a seat wrench or a valve seat removal tool, which is square on one end and hexagonal on the other. To remove and replace a valve seat, follow these steps:

➤ Remove the handle and stem, and insert the wrench squarely into the valve seat, turning counterclockwise.

➤ Loosen and remove the valve seat, examining for cracks or pits.

➤ Take the valve seat to your plumbing supply store, and buy an *exact* replacement.

➤ Place the new valve seat on the end of the seat wrench, apply pipe joint compound (pipe dope) to the threads, and carefully reinstall.

➤ Reassemble the faucet stem and handle.

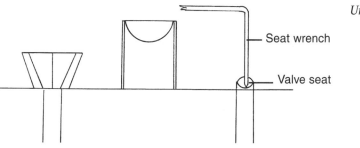

Seat wrench

Valve seat

Unseating your valve seat.

You might have an ancient faucet with a hard-to-replace valve seat or one that does not unscrew. In this case, you can grind the valve seat with a valve seat dresser, also called a faucet seat reamer, another handy tool for old plumbing. Insert the reamer inside the valve seat, and turn it clockwise a few times. Wipe out the shavings with a damp rag, and then run your finger inside the valve seat, checking for smoothness. You don't want to grind too much, just enough to smooth it out. Reassemble the faucet after the valve seat has been ground.

A valve seat gets introduced to a valve seat reamer.

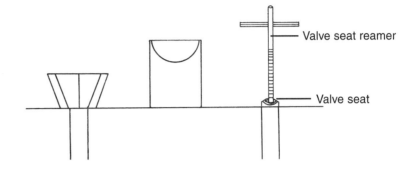

Valve seat reamer

Valve seat

Leaking Handles

Spouts aren't the only part of a faucet that can leak. Handles can also leak. A leaking handle suggests one of the following problems:

➤ A loose packing nut or locknut

➤ A damaged O-ring

➤ Deteriorated packing

Sometimes all it takes to stop a leaking handle is a slight tightening of the nut holding the stem in place. *Slight* can mean as little as an eighth of a turn to as much as a half turn. If the nut is already tight, that isn't the problem—you'll need to replace the O-ring or packing, whichever is present.

To replace a stem's packing or O-ring, remove the stem in the same manner described for replacing a washer. If your spindle has an O-ring, slip it off (you might have to pry it off carefully with the end of a knife) and roll a new O-ring onto the stem. Replace all washers and O-rings that are on the stem (the number varies depending on the faucet).

Older faucets won't have O-rings, but you will often find rope-style packing wrapped around the stem. Packing composition varies and can consist of carbon yarn and braided wire, graphite, and even Gore-Tex, depending on the application. Replace the old material

Plumbing Perils

A slow drip can waste as much as 20 gallons of water a day. A fast-dripping faucet is a lot worse. This really adds up over the course of a year and is a good incentive for repairing your leaking faucet.

with new packing, wrapping it around the same section of the stem. In a pinch, you can use Teflon tape, but packing should be available at a plumbing supplier.

Reinstall the stem and reassemble the faucet.

Washerless ... Sort Of

A washerless faucet uses one of the following to control the flow of water:

➤ A cartridge

➤ A (rotating) ball

➤ Ceramic discs

These mechanisms control water flow when their holes or ports are lined up in a specific manner over the water inlets. This is done without the pressure required in compression models. A washerless faucet doesn't work strictly without washers because it still requires plastic or neoprene material (an O-ring or gasket, for instance) between the cartridge or ball and the faucet seat. It simply doesn't compress (and wear down) a traditional rubber washer.

Each faucet has its own repair procedure, but they all start with the same first step: You have to shut off the water!

Repairing a Cartridge-Style Faucet

This faucet style's single handle controls water temperature and volume by having water run through the openings in a cartridge for the hot and cold water. The cartridge moves up and down to control water volume or flow, and it rotates to control the water temperature. A leak at the base of this faucet suggests a worn spout O-ring; a leak in the spout is due to a worn cartridge. Regardless of which one you're replacing, it's a good idea to replace all these components at the same time.

To disassemble, and repair a cartridge faucet, follow these steps:

➤ Shut off the water and lift the handle to drain any water out of the faucet.

➤ Remove the trim cap by prying carefully with a narrow screwdriver, and then remove the screw that secures the handle to the cartridge; lift the handle up and off.

Plumbing Perils

Be sure that you get the right replacement parts for your faucet. Write down the faucet's name and model before going to your hardware store or plumbing supplier. There are too many repair kits and parts to choose among and it's easy to get the wrong one. Buy a kit that includes an adjusting wrench and all seals, O-rings, and springs.

➤ Use a pair of channel-joint pliers to remove the retaining nut that holds the spout in place; move the spout back and forth as you lift it off.

➤ Find the small U-shaped retaining clip that secures the cartridge, and remove it with a needle-nose pliers (you may have to pry it slightly with a narrow screwdriver first).

➤ Pull out the cartridge with your pliers; if it sticks, place the handle back on and give it a few twists.

➤ Take the old cartridge to your hardware store or plumbing supplier, and buy a replacement. Install the new cartridge according to the manufacturer's instructions; it must be in the correct position, or it will reverse the directions in which the handle must be moved when you turn on the hot or cold water (hot will be on the right, and cold will be on the left).

➤ Install the retaining clip, and reassemble the faucet.

➤ Turn on the water, and check for leaks.

Sometimes a stiff cartridge might require a puller. Moen makes one for its cartridges that can be very helpful on stubborn, well-used cartridges.

While the faucet is disassembled, be sure to clean the spout and handle thoroughly with a mild soap and water.

All About Rotating-Ball Faucets

This faucet uses either a plastic or a metal ball with three openings to control the water. It can drip at the spout or at the base of the faucet. If you have to disassemble and repair one of these leaks, you might as well replace the necessary components to prevent the others from occurring. Replacement kits containing a new ball, seals, and springs are available, as are O-rings for the spout.

Pipe Dreams

If you can find a repair kit with a metal replacement ball, use it instead of plastic. The metal should last longer.

To repair a rotating-ball faucet, follow these steps:

➤ Turn off the water first.

➤ With an Allen wrench (which may come with the kit), loosen the set screw that secures the handle; stop before the screw is completely loose so that it stays in the handle (it's easy to lose).

➤ Remove the handle.

➤ Use the wrench that came with the faucet or that is included with the repair kit, and tighten the adjusting ring by turning the wrench clockwise (the ball should still move easily); install the handle and test for leaks.

➤ If tightening the adjusting ring doesn't stop the leak, wrap the serrated edge of the cap with tape, and remove the cap with a channel-joint pliers, turning counterclockwise.

➤ Note the position of the ball, and then remove it.

➤ Remove both the seals and the springs using a needle-nose pliers, if necessary; remove the spout O-rings.

➤ Take the old parts to your plumbing parts supplier, and get exact replacements (O-rings may be included in the repair kit).

➤ Install the new components according to the kit's instructions; note that the slot on the bottom of the ball must line up with the alignment pin on the bottom of the faucet's body and the cam with its corresponding notch.

➤ Reassemble and test faucet.

If your kit doesn't include an Allen wrench and none of your wrenches seem to fit, you might have a metric set screw and will need to purchase a metric Allen wrench.

Caring for Ceramic Disc Faucets

This third type of washerless faucet uses two hard, polished ceramic discs, one stationary and one moving, to control water flow and temperature. Sometimes leaks can be repaired by disassembling and cleaning the components rather than replacing them. You may or may not have an O-ring, depending on your specific faucet.

To repair a ceramic disc faucet, follow these steps:

➤ Shut off the water.

➤ Remove the screw securing the handle; this screw may be under a button cap, which will have to be pried off first.

➤ Remove the cap; it may lift off or have to be unscrewed in a counterclockwise direction.

➤ Loosen and remove the screws securing the cartridge; remove the cartridge and clean and rinse it thoroughly.

➤ Either replace the seals at the bottom of the cartridge, or replace the entire cartridge (use the correct replacement cartridge) and reinstall according to its instructions. Be sure that the ports or openings on the bottom of the cylinder line up properly with the faucet body.

➤ Reassemble the faucet, and test for leaks.

Plumbing Perils

Save any small tools that come with your washerless faucets. These are specifically designed to disassemble the faucet when it must be repaired. Store the tool in a clearly marked envelope in your kitchen or tool box.

Cartridge faucet.

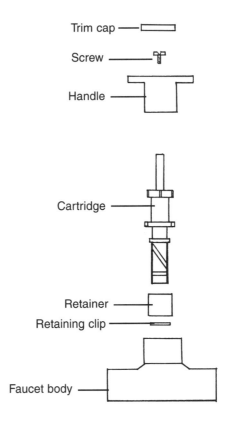

Trim cap

Screw

Handle

Cartridge

Retainer

Retaining clip

Faucet body

These faucets are pretty reliable, so check for grit or other contaminate between the disks before you run out for a replacement cartridge.

What's That Thingamajig?

A **clear coat** is a sealant such as a polyurethane applied to brass or other metal to retard tarnishing or oxidation. As long as the metal is sealed off from air and moisture, it will retain its shine. Once the clear coat wears away, the metal can begin to tarnish.

A Fitting Finish

Nickel-plated faucets and handles are long-lasting and have been used traditionally in kitchens and bathrooms because they don't tarnish. Chrome-plated brass is another excellent finish and will last forever with only modest cleaning requirements. Brass finishes have gotten more popular and come *clear-coated*, but they can be more problematic than nickel or chrome. Not every clear-coat finish lasts forever, though, and eventually that brass will need polishing. One advertised exception, however, is the Delta Brilliance line, which features a lifetime-guaranteed finish that is resistant to more than 100 common household cleaners. Regardless of the brand, follow the manufacturer's cleaning instructions and avoid abrasive cleaning agents.

190

Spray Attachments

Spray attachments are like spout extensions. Lift it up and out, and you can rinse anything, which is a real plus with double-bowl sinks. A sprayer is composed of a number of components that can deteriorate or become clogged from hard water deposits. These components can be cleaned or replaced, or the entire spray head can be replaced.

A definite sign of trouble is a decreased water flow from the spray head, in which case you should do the following:

➤ Check the hose under the sink to be sure that it isn't kinked; the hose can always be replaced if it's in bad shape by removing it from the underside of the faucet (this could require a basin wrench) and pulling it up through its hole in the faucet set.

➤ Turn off the water to the faucet, and remove the spray head by unscrewing it from the end of the hose.

➤ Either remove the screw cover, screw, perforated disk, and other parts, clean them with vinegar and a small brush and replace the washers, *or* simply replace the entire spray head.

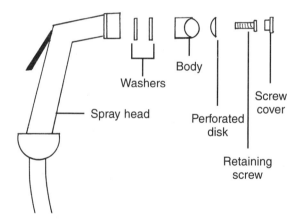

A typical kitchen sink sprayer.

The diverter valve, which directs water to the sprayer when it's in use, can become clogged or deteriorate and need replacement. In single-handle faucets, the valve is located behind the swivel spout, as shown in the diagram. If your spray hose checks out and you still have water flow problems, check the diverter valve next.

To remove and repair a diverter valve, follow these steps:

➤ Shut off the water and disassemble the faucet, removing the spout.

➤ Pull out the valve with a needle-nose pliers or your fingers, and clean it with vinegar and an old toothbrush.

➤ Replace the O-rings if they look worn, or simply replace the entire valve (take it with you to your plumbing parts supplier).

➤ Insert the cleaned or new diverter valve, reassemble the faucet, and turn on the water.

Check the prices on replacement spray heads and diverter valves as well as their individual components. It could be more worth your time to just replace them in their entirety rather than disassembling and replacing individual parts.

Plumbing the Depths

Vinegar consists mainly of acetic acid and water. The acidic qualities of vinegar are very handy when it comes to dissolving hard water deposits not only in faucets, but also in aerators and shower heads. It is a harmless, nontoxic solvent and an inexpensive fix when you need to remove these deposits.

Plumbing Perils

If bits of rust show up repeatedly in your aerator(s), and you haven't done any work on your own plumbing system, check with your water utility. They could be doing some work on the main or might not be aware that rust is being dislodged somewhere in the system.

Aerators

An aerator is a small screen that is screwed onto the end of sink and lavatory faucets to slow the flow of water. If aerators weren't installed, so much water would come out that it would splatter all over. You want the flow of water controlled in kitchen sinks and bathroom lavs. On the other hand, you want water to come gushing out in bathtubs where you want to draw the water quickly or in utility sinks where you want to fill a bucket or wash out a paint brush.

Aerators occasionally get clogged with hard water deposits or bits of rust that break loose somewhere in the water supply system. To clean an aerator, follow these steps:

➤ Unscrew the aerator by hand from the end of the spout; if it's necessary to use a pliers, wrap a rag around the aerator first to avoid scratching the finish.

➤ Rinse out any noticeable bits of rust or other debris.

➤ If you have hard water deposits, soak the aerator in vinegar and then rinse.

➤ If the aerator cannot be adequately cleaned, replace it and hand-tighten its replacement.

Sometimes a clogged aerator is the only problem with a slow running faucet. If only all your plumbing repairs could be this easy!

The Least You Need to Know

➤ Washerless faucets are the faucet of choice in kitchens and are generally quite reliable.

➤ The most common repair in a compression faucet is replacing a deteriorated washer.

➤ Always take your old parts or mechanisms with you when you go to your plumbing supplier for new faucet parts.

➤ Leaking faucets can waste a deceptively large amount of water if left ignored.

Chapter 14

Tubs and Showers

In This Chapter

➤ Detecting and fixing faucet leaks

➤ Unclogging clogged shower heads

➤ Diverter and spout repairs

➤ Disassembling different faucets

➤ Keeping tub surrounds sound

In the cinema, bathtubs and showers have a rich history. More than a few memorable scenes have taken place with one or more of the main characters bathing or showering (*Psycho* obviously comes to mind). To be able to bathe in warm water that's quickly available with few restrictions on how much we can use is really one of a plumbing system's great luxuries. Europeans may think that we bathe too frequently, but who wouldn't, given the unbelievable bathrooms we put together?

We've progressed from the weighty, cast-iron tubs from years ago to acrylic, enameled steel, and jetted tubs. Proprietary products such as American Standard's Americast Tub claim to be more durable than cast iron (which is still available) at half the weight while maintaining water temperature. Shower arrangements are available that can reproduce a misty rain or a waterfall. Single shower heads can be upgraded to models that offer variable sprays, including a massaging jet of water. No, it's not the same as the real thing, but it has its place.

Tubs and showers don't require a lot of maintenance. The faucets are subject to the same leaks and age-related ailments as a sink or lavatory faucet, and the drains can become clogged. Leaks around the tub are an issue unique to tubs because of all the cascading water coming from a shower head. The seal between the tub and the tub surround must maintain its integrity—otherwise, you can have big problems inside your walls.

This chapter covers normal tub and shower maintenance and repairs for a do-it-yourselfer.

Hide-and-Seek Shower Components

Shower valves and drains are hidden away in the walls and floor of your bathroom. As a result, they're not always user-friendly when it comes to repairs. Smart owners have their builders install access panels behind the tub in an adjoining room for future repairs. Smart architects will design the rooms so that these panels end up inside closets. Without a panel, you might have to cut into a wall to get at the valves or the ceiling below to get at a clogged drain pipe.

Problems that can occur in your tub and shower include these:

➤ Water dripping from the spout

➤ A weak shower spray

➤ A leaking faucet

➤ Water not properly diverted from the spout to the shower

➤ A leaking shower head

➤ Water leaking from the tub while the drain stopper is closed

These problems involve water trying to make its way into the fixture. Water going down the drain, the other category of tub and shower challenges, was covered in Chapter 10, "Clearing the Clogs." The figure shows the components of a typical shower.

Some older tub plumbing does not come with individual shutoff valves; those that do, even in new homes, aren't necessarily accessible unless a removable panel is provided. You're always safe turning off the water at the main shutoff to the entire system, even though that won't make you the most popular person in the household. Instead of jumping right into valve replacement, we'll start with the easy problems first.

Pipe Dreams

Consider installing a removable panel that would give access to your tub and shower plumbing before you need it, especially if you're going to be repainting the wall or doing other work in the adjoining room. This is an especially good idea if you have old plumbing.

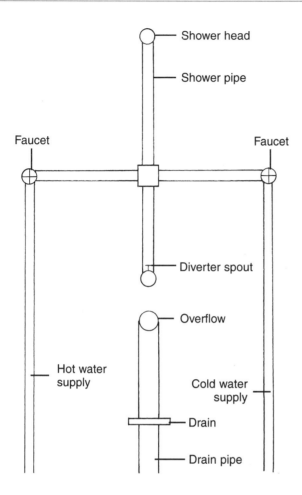

Typical shower and tub plumbing.

A Shower or a Dribble?

A shower head that has a gradually decreasing water spray probably is clogged with mineral deposits from hard water. There are two ways to dissolve these deposits:

1. Remove the shower head, pick away at the deposits, and soak it overnight in a bowl of vinegar.

2. Soak the head in vinegar without removing it.

A shower head tightens onto the shower arm—the pipe coming out of the wall—when turned clockwise. To loosen it, wrap a small rag or masking tape around the collar that secures the shower head to the shower arm. Turn counterclockwise. If you get a lot of resistance, place a second wrench or channel-joint pliers around the shower arm to give you more leverage and prevent it from turning (wrap the shower arm with tape or a small rag to prevent scratching the finish).

Once the shower head is removed, poke the spray holes with a toothpick or small-finish nail. Place the shower head in a bowl of vinegar overnight, and rinse with clear water the following day. When reinstalling, wrap the threads on the shower arm with a layer of Teflon tape, and tighten the shower head by turning clockwise.

You also can dissolve the deposits by filling a plastic bag with vinegar and placing it over the shower head without removing it from the shower arm. Secure the bag by wrapping it around the arm until the top of the bag is tight, and then wrap a piece of tape around it.

Plumbing the Depths

Old shower heads let loose with a virtual flood of water compared with new water-saving models. We stayed at The Mansions in San Francisco, an old Victorian inn, and the shower head was so big that the tub drain couldn't empty the water fast enough. From a water conservation standpoint, this obviously leaves a lot to be desired, but from a showering perspective, it's great. If you have an old shower head that's clogged with deposits, you might consider a new water-saving replacement.

Leaky Shower Head

This is another simple repair. First, determine where the water is leaking. Leaks at the point of attachment where the shower head screws onto the shower arm suggest a loose connection. Try tightening the head slightly. If you still have a slight drip, re-move the shower head, wrap some new Teflon tape around the arm's threads, and reattach. Tighten the head, but don't go reaming down on the wrench. Stop when you get reasonable resistance.

If you're getting drips out of the spray holes, you probably need a new washer. Remove the shower head and replace the washer (you may have to disassemble the head first by removing the screw that secures the faceplate). If this doesn't resolve the problem, your faucet could be the culprit.

Deciding Not to Divert

Some tub spouts contain a diverter valve that stops water from entering the spout and, as its name implies, diverts it to the shower pipe. If it fails to divert the water satisfactorily, replace the spout. Bathtub spouts are attached in one of two ways:

1. An Allen screw secures the spout to the nipple coming out of the wall.

2. The spout screws directly to the threaded nipple.

The Allen screw will be located on the bottom side of the spout. Loosen it and pull off the spout. If there is not an Allen screw, the spout itself needs to be turned counterclockwise. The easiest way to do this is by sticking a large screwdriver or the handle of a wrench into the end of the spout for leverage and then turning.

Take your old spout to your plumbing supplier, and buy an identical replacement. Install the new spout after applying Teflon tape to the threads of the nipple.

Plumbing Perils

Be sure to buy an identical spout to replace the old one you're removing. Otherwise, it won't fit properly or seal against the tub wall. In some cases, if an exact replacement isn't available, you'll have to remove the nipple and get a new one to accommodate the new spout.

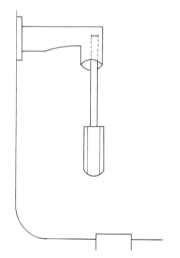

Unscrewing a tub spout.

Drain Stoppers That Don't Stop

Sometimes a drain stopper or mechanism can be defective or in need of cleaning. The simplest stopper to work on is the oldest: a rubber stopper that can easily be replaced. Other drain types, including the pop-up drain assembly, trip lever assembly, and the metal stopper with a drain flange, were covered in Chapter 10.

Leaky Faucets

Bathtub faucets are the same type, and of similar construction as sink and lav faucets, but the repairs aren't quite as simple. For one thing, the shutoff valves to the bathtub water supply aren't easily accessible unless you have an access panel. Second, disassembling the faucet isn't as easy as repairing one used in a sink or lav because the bathtub faucet stems are recessed inside the walls.

Two-handled bathtub faucets can be compression-style with washers, or cartridge-style with ceramic discs. Single-handle faucets can be either the ball-type, the disk valve type, or the cartridge-type. Each one is repaired differently, but all must be disassembled first.

Two-Handled Faucets

Hold your hand under the spout to determine whether the hot or cold water side of the faucet is leaking. Shut off the water to that individual faucet. If you have an access panel behind the bathtub, it will have individual shutoffs for the faucet. Other-wise, turn off the water at the main shutoff (give your family plenty of warning) and open the faucet to allow the water to drain out. Next, cover the bottom of the tub with a towel or drop cloth to protect it and catch any small parts that you might drop.

To disassemble a two-handled faucet, follow these steps:

➤ Pry off any decorative cap from the handle (use a small screwdriver or pocket knife).

➤ Remove the screw securing the handle; pull off the handle as well as the escutcheon or decorative plate behind it.

➤ Remove the stem or cartridge, depending on the type of faucet, with either a socket and a ratchet, or a special deep-set socket wrench sold for this purpose (see diagram); turn counterclockwise until the step or cartridge is removed.

➤ If necessary, *carefully* chip away any tile that may be obstructing access to the stem (just re-move enough to be able to use your wrench).

➤ Remove and replace all washers, O-rings, and any diaphragm; take them and the stem to your plumbing supplier to ensure exact replacement.

➤ Check the condition of the valve seat, and re-place it if necessary by removing it with a seat wrench (see Chapter 13, "Start with a Faucet"), or grind it with a valve-seat dresser to smooth it out; remove filings with a moist rag.

Pipe Dreams

It's always a good idea to take the miscellaneous components such as washers and stems to your plumbing supplier when you need to buy matching or compatible parts. This is no place for guesswork because it could cost you another trip to the store.

➤ Reassemble the faucet by installing the stem and turning clockwise. Test for leaks, and finish installing the escutcheon and handles.

If you're repairing a cartridge-style faucet, replace it with an identical new cartridge from your plumbing supplier.

Single-Lever Faucets

Single-lever bathtub faucets are similar to kitchen faucets, except that they're larger, of course, and have escutcheons. Each of these noncompression-style faucets has its own parts (a rotating ball assembly, a cartridge, or discs) that will require inspection and usually replacement.

To start your repairs, follow these steps:

➤ Shut off the water and drain the faucet.

➤ Remove the handle cover.

➤ Remove the screw securing the handle and then the handle itself.

➤ Remove the escutcheon.

➤ Remove and replace components as needed. (See the section "Washerless ... Sort Of," in Chapter 13.)

Plumbing Perils

Removing screws that have been in wet environments for years can be a challenge. Use the correct size of screwdriver so that you don't damage the head of the screw. Sometimes tightening the screw slightly will move it enough to loosen it. Consider heating the head with a propane torch (use only on metal handles, not plastic). Tapping the screwdriver lightly with a hammer can also help.

Plumbing the Depths

It's always easier to do your first tub faucet repair if you have another bathroom you can use in case you run into problems during your repair. If you're still a one-bathroom household and your plumbing is fairly old, you could go without a usable shower or tub over a weekend if you can't finish the work. Inspect the job first. You'll know soon enough whether you should proceed. If you call a plumber, be clear that it's an old faucet. Some plumbers specialize in maintaining old plumbing and carry lots of odd and weird parts in their trucks.

Rotating ball shower valve.

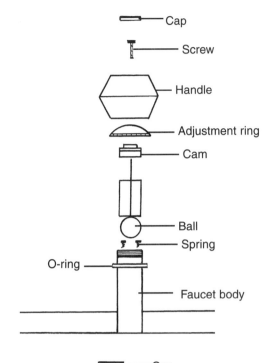

Cap

Screw

Handle

Adjustment ring

Cam

Ball

Spring

O-ring

Faucet body

Single-handle faucet with cartridge.

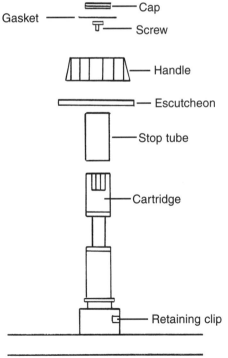

Cap

Gasket

Screw

Handle

Escutcheon

Stop tube

Cartridge

Retaining clip

Depending on the year of manufacture, your faucet might have been sold with a life-time guarantee. It's always worth inquiring at your plumbing supplier. If it is covered, you'll probably have to order the new parts directly from the manufacturer, in which case it might be faster to just buy them outright.

Tub Surround Integrity

Tubs and showers have more potential for leaking through the surrounding wall than sinks do through a counter. You might splash water around the edge of a sink that has a poor seal, but the water doesn't come down by the bucketful like it does in a shower. Tubs and showers can leak at a number of places, including these:

➤ At the joint between the tub and the surrounding wall

➤ At the joint between a shower base and the surrounding wall

➤ Through the tile grout

➤ At the joint between the tub or shower base and the floor

All these areas need to be inspected periodically for any separations or other breaks in the seals. Grout will have either small cracks or possibly entire sections missing. The other joints are typically sealed with silicone caulking that can pull away and need replacement. Both these materials need some attention from time to time.

To replace or maintain tile grout, follow these steps:

➤ Scrape out any deteriorated grout with an old screwdriver or knife blade; if there's any dampness in the joint, dry it with a hairdryer.

➤ Use either premixed latex grout or a powder grout mix in a color that closely matches your existing grout to replace the missing material.

➤ Wear a pair of latex gloves, and mix the grout with a small putty knife; force it into the joints with your finger or the putty knife (follow the directions that come with the grout).

➤ Wipe off any excess grout from the tiles with a clean, moist sponge; following the directions for drying time before doing a final wipe.

➤ Again following the instructions for drying times, seal the repaired area with a clear silicone sealant following its directions.

Plumbing Perils

Don't use regular latex caulking to seal a tub or shower: It isn't as long-lasting or protective in wet environments as caulking that contains silicone. Latex products are suitable for sealing gaps in wood trim and between the trim and drywall prior to painting.

If you've got badly deteriorated grout, especially near the edge of the tub or shower, you could have water damage behind the tile. This can be a real problem if the tile was installed over drywall instead of cement backerboard. Probe the wall behind the grout with a knife blade. If the blade goes all the way through, you might have to replace part of the wall or at least look into it further.

If you're uncertain when the tile was last sealed, go ahead and seal all the tile. It will have to be thoroughly cleaned first to remove any soap film or mildew (clean before you do any grout repairs). Allow plenty of time for the tile to dry before applying the sealant. Silicone sealant takes at least 24 hours to dry, so your shower will be out of commission during that time.

Tile Replacement

This can be a tricky job. If you're not careful removing a cracked tile, you can damage an adjoining tile. You can always skip the job for a while and run a thin bead of clear caulking in the crack to prevent water from leaking behind the tile.

To remove a cracked tile, follow these steps:

➤ Place a drop cloth or piece of plastic in the bottom of the tub or shower stall.

➤ Tape the edges of the surrounding tiles with two layers of masking tape (this will protect their edges when you chip out the other tile).

➤ Place a piece of masking tape in the center of the tile to be removed, and drill through it with a masonry bit (the tape will keep the bits of tile from flying around too much).

➤ Drill just enough to get through the tile (you don't have to go through the wall); the more holes you drill, the less chiseling you'll need to do.

➤ Use a cold chisel and hammer, and gently break the tile out.

➤ Use the chisel or a stiff putty knife to scrape out any remaining grout and adhesive before installing the new tile.

➤ Apply tile adhesive to the new tile with a putty knife (use a serrated trowel if the back of the tile is flat or smooth; if it's ribbed, use a smooth putty knife).

➤ Set the tile in place, press firmly, and insert spacers (available at your tile supplier) on all four sides to ensure even spacing.

➤ Use masking tape to secure the tile overnight while the adhesive dries.

➤ Grout the tile after its adhesive has dried.

If you cannot find a matching tile, consider a one-of-a-kind artistic tile or a contrasting color.

Pipe Dreams

Always keep some extra tile in storage from any tiling job in your house. It could be years before one needs to be replaced (the next owner may be the one doing the replacement), and you'll never be able to match the tile later.

No Tile Problems with Acrylic

One-piece acrylic tub and shower units solve all kinds of leakage problems because they don't leak! For as little as $200—less than the cost of the tile alone, in most cases—you can have a finished unit that's easy to clean and maintain. No seams, no leaks at the walls. The only joint you'll have to think about is where the tub meets the floor. If you have an acrylic unit, follow the manufacturer's directions for cleaning and coating with special products made to maintain the materials finish.

Laminated Surrounds

You might have an existing tub that you really want to keep, but the walls around it might leave something to be desired. An alternative to tile is a laminated tub surround. This is a single sheet of plastic laminate glued to the walls. There are no seams except where the laminate meets the tub. You have a wide choice of laminate colors and patterns to choose from, and you can match the laminate used on your vanity top.

Installing a plastic laminate tub surround is best left to a professional who specializes in this type of work. The laminate does not have to be installed over cement backerboard like tile because it's a solid material without any grouted areas that can leak water. It does require a smooth wall for installation, though.

Be sure to check with your installer about warranties for the laminate itself and the installation.

Pipe Dreams

The best way to maintain an as-new appearance of any tub or shower surrounds is by wiping them dry after each showering. Keep a squeegee in the shower to give the walls a quick swipe. Tile can go for years without resealing if it's kept dry this way. Laminate, which can easily show water stains, especially benefits.

The Least You Need to Know

➤ Tub and shower faucets and spouts function the same as those in sinks, but they're more complicated to disassemble and repair.

➤ Look for the simple solutions first—such as tightening a shower head that's leaking—before you really start tearing into things.

➤ Always protect the tub itself by covering the bottom of it with a drop cloth or towel while you work on the faucet, spout, or shower head.

➤ There's no substituting the real thing when buying replacement parts for your faucets; take the old parts with you to the plumbing supplier when buying new.

➤ With a little regular maintenance, your tub surround can last for years without any leaks.

Part 4

Bigger Fixes

With the exception of those constructed with certain varieties of flexible plastic pipe, most plumbing systems will be long-lasting and mostly trouble-free. Of course, you might be dissatisfied with the vintage and style of some of your fixtures and want to replace them. Plumbing upgrades might be part of a general remodeling, such as installing a new kitchen (time for that single-bowl sink to go).

Replacing fixtures can be a bit of a chore. The old sink, tub, or toilet has to be removed and discarded; all the old caulk and sealant must be cleaned away; and the new fixture must be hauled in and finally installed. Tub and shower installations are especially time-consuming and messy, but it's worth the trouble if you just can't stand staring at that turquoise porcelain another day.

Kitchens offer all kinds of opportunities for change and upgrades. Older homes came with a sink and hot- and cold-water taps, and nothing more. This won't do for most modern families. Now, in addition to the sink, you can add a dishwasher, a disposer, a second or even third sink, an instant hot-water dispenser, and a refrigerator with an icemaker. We'll discuss all these options in Chapter 17, "Kitchen Concerns."

Chapter 15

Replacing Sinks and Faucets

In This Chapter

➤ Self-rimming and rimless sinks

➤ Installing stop valves

➤ Buying the right faucet

➤ Disconnecting everything below

➤ Hooking up the pipes

It's not every day that you need, or want, to replace a sink or a faucet. Sinks don't simply wear out one day and give up the ghost, but they do age and become stained and chipped, or they just go out of fashion. If you're remodeling a kitchen or bathroom, you'll almost inevitably end up replacing the sink and other fixtures.

When the sink goes, the faucet usually goes with it. What's the point of pulling out a 20-year-old sink but keeping a 20-year-old faucet? In fact, you probably should pick out the faucet first and then find a sink to match its configuration. Sinks come pre-drilled for faucets and, in the case of kitchen sinks, accessories such as sprayers and soap dispensers. It's possible to drill additional holes in an existing sink, but you're better off with one that's ready to go as it is.

Use the sink and faucet you're replacing as guides to their replacements. Note what you don't like about them so that you'll be sure to avoid these features when you look at new fixtures. One walk through a plumbing showroom or home improvement store will show you that you have no shortage of choices.

First, a Look on Top

Most built-in kitchen and bathroom sinks come in two styles:

➤ Self-rimming

➤ Flush-mounted

Pipe Dreams

If you plan to replace a stainless steel sink with a new model because your old one looks dull beyond repair, try some cleaners and even an electric buffer first to see whether there's any life left in it. It's always easier to keep an existing sink if it's salvageable rather than to replace it.

A self-rimming sink is the easiest to install because its rim rests on the counter itself, just outside the perimeter of the hole into which the sink fits. This gives plenty of support and lets you fudge the fitting a bit if your hole is a little off in its dimensions. The drawback to self-rimming sinks—at least with cast-iron and composite models—is the height of the rim relative to the counter. You can't exactly sweep crumbs and celery leaves off the counter and into the sink with this rim in the way. Still, these are popular and easy to install.

A flush-mounted sink is rimless and attaches to the counter via a rim that is mounted independently of the sink. The sink is then attached to the rim with a series of clips. These sinks allow for easier cleaning of the counters but are more difficult to install.

Stainless steel sinks straddle both worlds. They are always self-rimming, but their rims are so flat that they are easy to clean around.

Looking Below

Take a look under your sink—I mean, really under it, not just in the cabinet. Shine a flashlight up and see how your sink is secured to the counter. A self-rimming cast-iron sink will rest on its own weight. It doesn't need any additional mechanical attachments to secure it.

Plumbing Perils

Stop valves aren't simply a convenience in the event of repairs or maintenance to your sink and faucet. You also want them in case you have a bad leak that's allowing water to run over a counter and onto the floor or inside a cabinet. A stop valve will help keep the mess to a minimum.

Lighter bathroom lavatories or stainless steel kitchen sinks are secured with a series of bolts or clips that are screwed to the underside of the counter. These must be removed before the sink can be pulled out. Flush-mounted sinks also have clips that must be removed.

Look for stop valves (these are shutoff valves that control the water flow to a single fixture). If you don't have these, you should install them as part of your sink or faucet replacement. You don't want to be shutting off all the water in the house if you have to repair the faucet at some point in the future.

Stop Valves

To install a stop valve, determine the kind of pipe you have (steel, copper, or plastic) and how it connects to your sink. In all cases, you should plan to replace the supply tubes that carry water to the faucet with new flexible tubes. Follow these steps to install a stop valve to steel pipe:

➤ Shut off the water and drain it from the faucet.

➤ Look for the fitting to which the supply tube is connected. (The pipe itself will come out of either the wall or the floor.)

➤ Disconnect the supply tube from the fitting (use an adjustable wrench).

➤ If your pipe comes out from the wall, install a half-inch by three-eighths-inch angle stop (this is designed to accommodate right-angle installations) to the threaded end of the *stubout* using pipe dope.

➤ Tighten the valve by hand, and then tighten with an adjustable wrench, making sure that the valve outlet to which the supply tube connects is pointing up.

➤ Install the supply tube (shown in the figure).

➤ Test the valve for leaks.

If your steel pipe comes up from the floor, install a straight stop instead of an angle stop.

What's That Thingamajig?

A **stubout** is the short section of pipe that extends from a wall under the sink. Stop valves or supply tubes connect to the stubout.

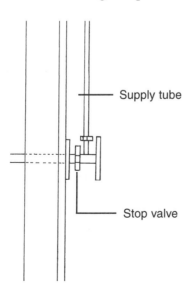

Supply tube

Stop valve

A typical stop valve and supply tube.

Copper and Plastic Pipe

Newer homes with copper or plastic pipe should already have stop valves installed, but that isn't always the case. It's one thing to excuse it in an old house, but there's no excuse in more current construction for not installing one of these valves. It's most likely that your pipe will be coming out of the wall and then angle up toward the faucet via an elbow fitting, but not always. The copper pipes could be coming up from the bottom of the cabinet or floor, such as in the case of an island sink.

If your pipes come out of the wall, follow these steps:

➤ Shut off the water and drain it from the faucet.

➤ Using a tube cutter or hacksaw, cut the pipe just short of the fitting. Make sure that you'll leave enough room to attach the stop valve.

➤ Clean the pipe and either solder on a threaded male adapter and then screw a threaded stop valve to the adapter, or simply attach a compression-style valve that is secured by tightening the compression nut.

Plastic pipe also will accept a compression fitting, or you can solvent-weld a threaded fitting to the stubout.

Installing stop valves is important and should take place right after your old sink is pulled out.

Removing Your Old Sink

Kitchen sinks are more difficult to remove than bathroom lavs because there are usually more encumbrances, such as disposers and dishwasher attachments. Plus, cast-iron double-bowl sinks are heavy and really need two people to remove them safely. Stainless steel sinks, on the other hand, are very lightweight and easy to pull out.

Follow these initial steps when removing any kind of sink:

➤ Clean everything out from under the sink—you'll be surprised at what you'll find and how much you'll toss out.

➤ Shut off the water and open the faucet to drain out any water.

➤ Place a pan under the sink, and disconnect the trap and tail piece using channel-joint pliers.

➤ Disconnect the water supply tubes from the faucet; if you're going to replace them, disconnect them from the stop valves or stubouts instead.

With the plumbing detached, the sink can be removed.

Removing the Kitchen Sink and Disposer

One item a kitchen sink often comes with that a bathroom lav does not is a food disposer. It's easier to remove the sink if you remove the disposer first. You should only have to loosen it and let it rest on an overturned bucket or wastebasket instead of completely removing it from inside the cabinet. To loosen the disposer, follow these steps:

➤ Unplug it or shut off the power. If the disposer is hardwired, the electrical cable should be long enough that you won't have to disconnect it from the junction box.

➤ Disconnect the disposer's drainpipe from the waste pipe and the dishwasher drain hose.

➤ If the disposer is held in with bolts or screws, loosen them while keeping one hand on the unit; lower the disposer as it drops down.

➤ If you do not see any bolts or screws, the disposer is held in by a support or retaining ring. Again, hold the disposer with one hand and turn the ring counterclockwise, using a screwdriver, if necessary, for leverage.

➤ Lower the disposer and put aside.

After the disposer is removed, look for any tabs that might be securing the sink. If you have a self-rimming sink, go ahead and unscrew the clips. With a putty knife, gently break the caulking between the edge of the sink and the counter.

Push up on the sink from underneath. If you have a heavy cast-iron sink, enlist a second person for help. As the sink lifts away from the counter, slip in a wedge of wood to prevent the sink from resting back on the counter.

Rimless or flush-mounted sinks can be lifted out or sometimes lowered once the securing clips are removed. Corroded clips and heavy, cast-iron sinks are a bad combination because the sink can actually fall through the opening as the clips are removed. To secure your sink and prevent this from happening, lay a two-by-four piece of lumber (or a scrap

Pipe Dreams

Unless they're in good shape, go ahead and replace your old trap and drain pipes. This is a particularly good idea with chromed pipe because it tends to corrode, while new plastic does not. Replacement parts are not a big-budget item.

Plumbing Perils

Don't be too aggressive breaking the caulking bead around the perimeter of a sink. The knife can slip, and you can end up scratching your counter. Ease the knife in, and slowly cut through the caulking while prying up against the sink at the same time.

piece of heavy metal pipe) that is at least a foot longer than the sink across the bowl(s) so that its ends rest on the counter. Using lengths of rope or bailing wire, attach this board to a second one running underneath the drain holes. If the sink does slip while you're removing the clips, it won't fall on you. Lift the sink up and out or, if possible, lower it and remove it.

Plumbing the Depths

If you want to install a larger sink than the one you're removing, you'll need room under the counter to do so. Measure the new sink, including its depth, and check for space. Use your new sink as a template (unless a paper template comes with it) for cutting out a larger hole in the counter with a saber saw. Trace the sink around its edge, and then draw another outline a half-inch inside the first one. Test-fit the sink when you're finished cutting. You'll need either a new counter or a repaired and resurfaced one if you want to install a smaller or differently shaped sink.

When the sink is removed, thoroughly clean any old caulking or putty off the counter before installing your new sink.

Removing Bathroom Lavatories

You won't find a dishwasher or disposer in your bathroom (okay, it's pretty unlikely, but you never know these days), so removing a bathroom lav is pretty simple. After disconnecting all the supply pipes and trap, gently cut through the bead of caulking sealing it to the counter. If you have a self-rimming lav, push it out from below, lift it out, and clean off the old caulking.

Some bathroom lavs are rimless and will be held in with metal clips. Use the same precautions here as you would with a kitchen sink. Either secure the sink from underneath as you undo the clips, or run a two-by-four across the top and tie it to a scrap of wood below the drain hole.

Sink and Faucet Selections

We discussed fixtures and faucets in Chapter 9, "Choosing Your Fixtures and Faucets," but now you've actually got immediate motivation to do some serious shopping. You know the benefits of one material over the other (porcelain vs. stainless steel, for instance) and can choose according to your needs. Be sure that the sink holes match up with your chosen faucet! Most faucets have standard measurements, but check against the spacing of the sink holes anyway.

When choosing a kitchen faucet, you'll have to consider these factors:

➤ One handle or two

➤ The type of hose sprayer

➤ The height of the spout

➤ The faucet's style

The most convenient spout, even for single-bowl sinks, is one that swings rather than one that's fixed. A high spout allows you to more easily fill tall pots and pitchers. The traditional sprayer and hose are separate from the spout, but an increasingly popular style is a pullout spout that works as a sprayer.

Installing Your New Sink

Regardless of whether you're installing a kitchen sink or a bathroom lav (or a scrub sink, for that matter), the procedures are similar. The easiest approach is to install as much of the pipes and faucet as you can before installing the sink itself. This way you're not spending any more time than you have to crawling around underneath the sink.

By now you've installed your stop valves, cleaned the old caulk and gunk off the opening in the counter, and have purchased your sink, faucet, and *waste kit*. A kitchen sink also will need a new strainer.

The strainer is the basket that attaches to the drain hole. Its purpose is to prevent large items from going down the drain and to act as a point of insertion for a stopper. If you have a double-bowl sink and are installing a disposer, which comes with its own hardware for the drain hole, then you'll need only one strainer for the other bowl.

You can install the following components before installing the sink:

Plumbing Perils

Shop for quality, not price. An inexpensive, off-brand faucet will not be as long lasting as a costlier brand-name faucet. You don't have to spend a fortune, but figure on $100 and up for a decent faucet that will last.

What's That Thingamajig?

A **waste kit** is a drain assembly package that contains the components you need to reassemble your drain pipe. These are universal components with $1\frac{1}{2}$-inch traps.

➤ The faucet

➤ The strainer

➤ The tail piece

The faucet is installed first because you want to be sure that there aren't any fitting or compatibility problems between it and the sink.

Installing Your New Faucet

If you're installing a new faucet, you don't have to worry about removing and reusing the old one whose fastening nuts and bolts might be rusted or corroded. (See "Removing an Old Faucet" later in this chapter.) Every new faucet comes with installation instructions, but we'll go over the general rules of conduct here. We'll start with the kitchen sink.

Place a towel or old blanket on your counter, and then place your new sink on top. This will prevent any unintentional scratches on your laminate or other counter material. Take out your new faucet set, and spread out the parts, comparing them against the checklist included in the instructions. You don't want to get halfway into the installation only to find that a part is missing.

Some faucet sets come with a built-in sprayer. Others do not, and the sprayer is installed separately in its own hole in the sink. The faucet set itself will be inserted into separate holes.

Your faucet will seal itself to the sink using one of these three materials:

Pipe Dreams

Many faucets come with their own supply tubes attached. You attach a second set of supply tubes that also attach to the stop valves. These faucets are very convenient and are well worth installing if they're available in a model and style that you want.

➤ A gasket (included with some faucet)

➤ Plumber's putty

➤ Silicone caulk

These materials are used to fill the gap that inevitably forms between the faucet and the surface of the sink. If your faucet comes with a gasket, use it and it alone. Otherwise, either the putty or the caulk will do the trick. Roll the plumber's putty between your hands until it's soft and pliable, and then roll it into long ropes (think of preschool days). Place the putty on the bottom of the faucet. (Don't use plumber's putty on a marble counter, however, because the marble can stain and crack.) Do the same if you use caulk by squeezing a bead of it from the tube.

Align the faucet with the sink holes, and press it into the sealant. Attach and tighten the fastening nuts and washers onto the threaded tail pieces of the faucet underneath the sink. When they're hand-tight, check again that the faucet is aligned properly with the back of the sink (it should be parallel, of course), and then further tighten the nuts with a basin wrench. Plastic nuts can be tightened by hand and then a final part twist with pliers. After the faucet has been secured, remove any excess putty or caulk from underneath the faucet.

More Preinstallations

It's likely that your kitchen sink set will come with a sprayer, so this can be installed next. Read the instructions that accompany the sprayer, which should echo the following steps:

➤ Insert the hose guide into the hole, and feed the hose through the guide.

➤ Insert the threaded tail piece into the hole.

➤ Secure the tail piece with the nut and washer that came with the kit.

➤ Attach the other end of the hose to the hose stubout (sometimes called a spray outlet shank) that is directly under the base of the spout.

You'll see how much easier it is to do these installations before the sink itself is set in place. The next piece to install is the strainer. There are various models with different installation instructions (read them first), but the following directions hold true for most strainers:

➤ Wipe the drain hole clean.

➤ Apply a half-inch roll of plumber's putty or a thick bead of silicone caulk to the underside lip of the strainer.

➤ Press the strainer into the drain hole.

➤ Insert the gaskets that come with the strainer onto the threaded neck that's now sticking out underneath the sink.

➤ Attach and tighten the lock nut using a pair of channel-joint pliers.

➤ Wipe away any excess putty or caulk from the inside of the sink (you want a small bead of it to remain between the strainer and the sink to act as a seal), being careful not to scratch the surface of the sink.

Install the mounting assembly for the garbage disposer, but don't install the disposer itself. This will be installed after the sink is in place.

Finally, install the tail piece that attaches to the strainer. You're finally ready to install the sink.

Preinstallations make for an easier sink installation.

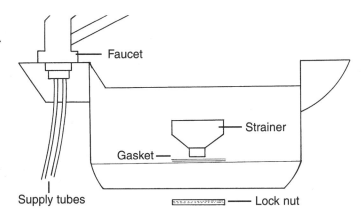

Installing a Self-Rimming Kitchen Sink

Self-rimming sinks are easier to install than rimless sinks. Follow these steps:

➤ *Test fit* the sink.

➤ Remove the sink, and run a bead of silicone caulk or a rope of plumber's putty around the underside of the rim's edge.

➤ Insert the sink back into the opening, and press down.

➤ Insert any mounting clips under the sink, and tighten the screws.

➤ Clean up any excess caulk or putty that squeezes out from under the rim.

Heavy cast-iron sinks, which often take two people to install comfortably, will not have any mounting clips.

What's That Thingamajig?

For a neater job when using caulk, **test fit** the sink into the hole and apply masking tape just outside the rim's edge. After the caulked sink has set for about 10 minutes, remove the tape (you don't want the caulk to completely dry because the tape can pull it off).

Installing a Flush-Mounted Sink

It's more likely that you'll be installing a self-rimming sink for the same reason that contractors do in new housing: They're simple, attractive, and easy. A flush-mounted or rimless sink requires a separate sink rim to be installed along with the sink. The two are secured with tabs and rim clips. To install a flush-mounted sink, follow these steps:

➤ Insert all the screws into the clips, and give them a turn or two so that they won't fall out.

➤ Match the rim with the sink, and bend all the rim tabs inward to secure it to the sink.

➤ Insert the sink into the counter opening (heavy sinks will need to be supported from below).

➤ Crawl under the sink and install all the clips (see the manufacturer's instructions for spacing recommendations).

➤ Tighten the clip screws.

As you tighten the clips, check that the sink is aligned properly and has remained parallel to the edge of the counter.

The World Below

Now it's time for the hookups, and your first one will be connecting your new supply tubes to the faucet and stop valves. The type of supply tube will determine its installation. A braided stainless steel tube is attached by simply tightening the coupling nut and washer to the faucet tail piece or supply inlet, depending on its design. The other end of the tube then is tightened to the stop valve. These are compression fittings and should be tight, but not excessively tightened.

To install plastic or soft copper supply tubes, follow these steps:

➤ Insert the end of a tube with the attached head into the valve tail piece, and hand-tighten it with the coupling nut (included with the faucet).

➤ Put the other end against the stop valve, and measure enough so that the end will be completely inside the valve.

➤ Mark this length with a laundry marker or piece of masking tape, and remove the supply tube from the faucet.

➤ Cut the tube with an appropriate cutter (a knife for plastic, a tube cutter for copper).

➤ Repeat for the other supply tube.

➤ Reinstall the tubes at the faucet, hand-tightening the coupling nuts.

➤ Install the coupling nut and any washer or ring that comes with it to the stop valve, and hand-tighten.

➤ Tighten the faucet end of the tube with a basin wrench and the stop valve end with an adjustable wrench until snug.

Pipe Dreams

It's a lot easier to work under a sink if you have some light. Flashlights are okay, but an electric work light is much more convenient. Try a Craftsman Professional 1250-watt work light. It comes with a magnetic clip as well as a hook for hanging it up and out of the way. It also has a 10-amp capacity outlet on the light itself for plugging in small tools.

Plumbing Perils

Flexible supply tubes come in varying lengths. Plastic ones can be cut to size, but the metal mesh versions cannot. Measure your old ones and purchase something comparable. You can always go a little long, but don't try to go shorter.

The world under your kitchen sink.

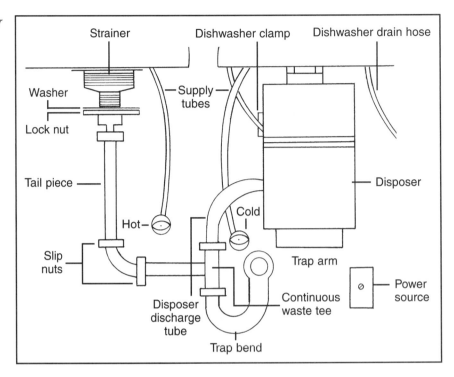

A second view under your kitchen sink.

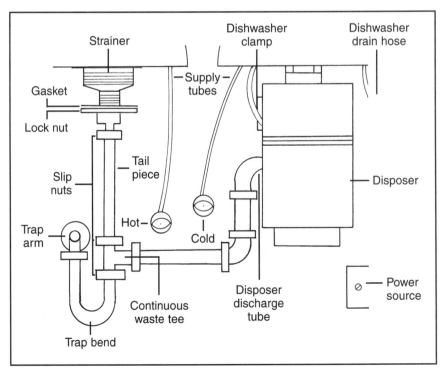

Installing Your New Disposer

With the supply tubes out of the way, install the disposer according to the manufacturer's instructions or by reversing the steps you used to remove it from the old sink. If part of your dishwasher drain hose was first connected to a counter-mounted air gap, reconnect this before you install the disposer (it's easier with the disposer out of the way). You might not have an air gap if your local code doesn't require it.

With the disposer installed, connect the dishwasher drain hose to the disposer by tightening the hose clamp.

You're almost done. All that's left is the drain pipe assembly.

Disposer Drain Pipes

The disposer has its own discharge tube that carries waste to the drain pipes. Depending on your arrangement, it will connect to either a continuous waste tee or a trap arm. The diagrams that follow show a couple different scenarios. After fitting all the sections of plastic pipe together, tighten the coupling nuts by hand. Gently tighten them further with channel-joint pliers. Be sure that all the threads line up with each other so that you don't strip them during the installation.

Pipe Dreams

The ideal time to replace a disposer is when you're replacing the sink. If your disposer is old or underpowered, plan to install a new one.

Testing, Testing

Now it's time to turn the water back on. Look for leaks at all the connections. Often these can be corrected by slightly tightening the coupling nut. When you're satisfied that the installation is leak-free, turn on the water at the faucet and check for leaks at the faucet itself, the strainer, and the drain pipes.

Bathroom Lavatories

A kitchen sink is definitely tougher to remove and install than a bathroom lav. The former is larger and often heavier, and has more attachments to deal with (dishwasher, disposer, sprayer). A bathroom lav is mounted either in a vanity or on a counter, or it's wall-mounted on brackets, with or without front legs for additional support.

A bathroom lav is removed the same way a kitchen sink is removed: Disconnect the water supply tubes and drain line, break any caulking bead between the vanity top and the lav, and gradually pry and lift the lav up and out. Be sure that your new sink fits the dimensions of your existing hole unless you can go larger and increase the hole's size.

Note the spacing of your new lav's faucet holes. As you know by now, they must match up with any new faucet you plan to install. Like kitchen faucets, bathroom models offer many design and style options, including these:

➤ One or two handles

➤ Various spout lengths and heights

➤ Different finishes

Pipe Dreams

Always put an old towel under the lav first while you work on it. You don't want to scratch it or your floor or counter.

Be sure that the faucet you choose not only fits the lav's holes, but also fits the style and overall size of the lav. Some might call out for a tall, graceful spout, while others would look best with a long spout that shoots out farther into the lav. Some handles are roomier and easier to clean than others, and this, too, is a consideration.

Installing a New Lavatory

With the old lav removed, clean the old caulking or plumber's putty from the counter. Clean up under the old faucet, too, if it was mounted on the vanity top and not on the sink. You'll want to install as many components as possible before installing the sink itself, mainly the faucet, the pop-up assembly, and the tail piece.

Most lavatory faucets are top-mounted models—that is, those in which the components are contained in an enclosed body that sits on top of the lav. The opposite model is a bottom-mount faucet that has only the spout and the handles mounted on top of the lav, while its working parts are underneath.

The pop-up assembly is the lav drain. After it's installed, the tail piece is attached. The assembly will come with its own directions (naturally). The diagram that follows shows the major parts of a pop-up drain assembly.

Installing the Lavatory Faucet

One difference between kitchen self-rimming sinks and those in bathroom lavs is the room available for mounting a faucet. It might be necessary to drill holes in the top of your vanity to accommodate the faucet. Drop the sink into the hole first, line up your faucet, and mark the holes so that you'll know where to drill.

Regardless of the faucet's location (on top of the sink or the vanity counter), turn the faucet upside down and insert any gasket that comes with it. If there isn't a gasket, apply a modest amount of plumber's putty or silicone caulk to the base of the faucet (again, do not use plumber's putty if the vanity surface or sink is marble). Insert the faucet into the sink holes, and hand-tighten the washers and locknuts onto the threaded tail pieces. Check that the faucet is aligned properly with the edge of the sink, and finish tightening the nuts with a basin wrench.

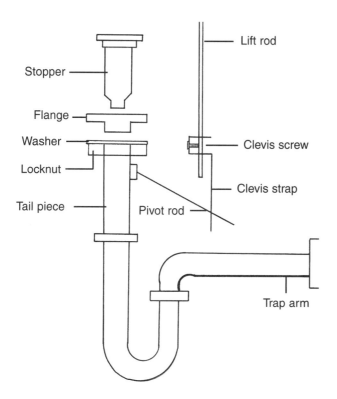

Stopper

Flange

Washer

Locknut

Tail piece

Pivot rod

Lift rod

Clevis screw

Clevis strap

Trap arm

Pop-up drain assembly.

After the faucet and sink are in place, hook up the water supply tubes and the drain pipes, which consist of the following:

➤ Tail piece

➤ P-trap

➤ Drain arm

➤ Slip nuts

The tail piece is the short vertical section that connects directly to the drain body right as it comes out of the drain hole in the sink. Next comes the P-trap and then the drain arm, which connects the P-trap to the stubout protruding from the wall. Assemble the sections in this order and along with the slip nuts. Hand tighten the slip nuts and then gently tighten with slip-joint pliers.

Turn on the water and test for leaks.

Drain pipe assembly.

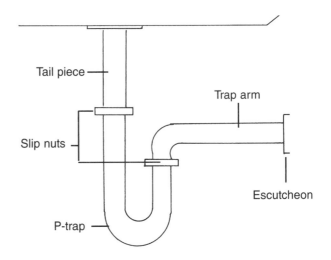

Tail piece

Trap arm

Slip nuts

Escutcheon

P-trap

Removing an Old Faucet

You might only want to remove an old faucet and replace it with a newer mode. Sometimes this is the only choice on really ancient faucets when fitting new parts is a problem. Old faucets can be a little cranky it comes to removal. Most older faucets will remove in the following way:

➤ Shut off the water and open the faucet.

➤ With a basin wrench, unscrew the nuts that secure the hot and cold water supply tubes; if these are soft copper tubes, gently put them aside unless you're going to replace them.

➤ Remove the nuts from the faucet's tail pieces (the threaded sections of pipe that fit inside the faucet holes in the sink).

Plumbing Perils

You can use a propane torch to loosen nuts if they absolutely won't budge, but be careful! Use a low flame, and keep a spray bottle of water nearby in case anything ignites.

If any of the nuts are corroded and have frozen, spray them with a penetrating lubricant and try them again after it has soaked in.

Some nuts can be loosened with other tools if you have enough room to fit them between the sink and the back wall. If your faucet has a pop-up drain, you'll end up replacing it as well. To remove the pop-up assembly, follow these steps:

➤ Undo any nuts or screws that secure the lift rod to the clevis strap and remove the rod.

➤ Loosen the retaining nut that holds the pivot rod inside the drain body, and remove this rod.

➤ Remove the stopper.

➤ Disassemble the p-trap, tailpiece, and the lower drain body (the section threaded and attached to the drain flange)

➤ Carefully push up and remove the drain flange and clean out any putty or caulking

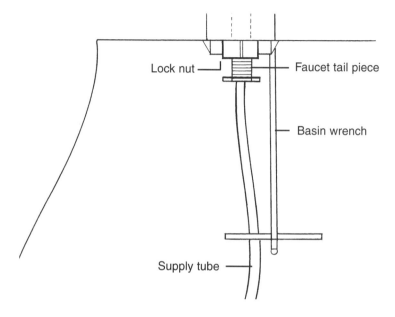

Removing a faucet.

Lock nut

Faucet tail piece

Basin wrench

Supply tube

Your new pop-up assembly should come with a new flange, lower drain body, and any gaskets or washers for a complete installation.

The Least You Need to Know

➤ Your first order of business when replacing a sink should be to install stop valves if you don't currently have them.

➤ Every sink is different when you try to remove it; look carefully underneath for any fasteners, and beware that cast–iron sinks are heavy and best removed by two people.

➤ When you replace your sink and faucet, shop around for the faucet first and be sure that your sink has the proper spacing in the faucet holes.

➤ Always preinstall as many components of the sink as you can before installing the sink itself; this makes for a much easier job.

➤ When all else fails, follow the directions that come with your sink, faucet, strainer, and disposer for proper installations.

Chapter 16

Installing a New Tub and Shower

In This Chapter

➤ Deciding when to replace

➤ Removing your old tub

➤ Rebuilding your walls

➤ Multiple shower heads

➤ Related installations

Many new master bathrooms come with separate tub and shower areas. This doesn't present any special plumbing challenges, but it does require two sets of drain pipes and vents and water supply pipes running to both fixtures. Installing these fixtures in a new bathroom is considerably easier than working in an existing bathroom. Old cast-iron tubs won't leave without a struggle, and that usually means taking a sledge-hammer to them. Even old fiberglass units can be a nuisance to remove.

In new construction, a tub or tub/shower combination is installed while the bathroom is still under construction. The tub is installed against the framing studs. The drywall or cement backerboard then is brought down to cover the lip of the tub. The seam gets sealed with caulk after the wall covering—tile, laminate, or other material—is installed. The tub is shipped in a cardboard box that can double as a protective cover after the tub is installed, but you should take additional measures to keep the tub from getting scratched.

We've already discussed the different types of tubs. (See Chapter 5, "Your Wish List.") The plumbing issues remain the same regardless of which tub or tub/shower you choose. You'll need access to your walls and floor to run branch pipes, drains, and vents. This chapter takes you through the process and also offers some alternatives to complete replacement.

Replace or Not Replace?

All a bathtub *has* to do is hold water, allow it to drain, and not leak. We could get by with an empty 50-gallon oil drum (cleaned first, of course) with a rubber stopper drain in the bottom on the end of a long cord or small chain. The drum could be filled with a couple washing machine hoses attached to a hot and a cold hose bib. You could live with this, but only if you like survivalist chic. Most of us want something a little fancier.

Pipe Dreams

To protect your tub during construction of your bathroom, cut a piece of plywood to fit over the top of the tub. Line one side of the plywood with an old blanket. Line the tub with a sheet of plastic, and then lay the plywood over it. This will both protect the tub and allow you or another worker to stand on it, if necessary.

There are various reasons for replacing an existing tub, including these:

➤ Outdated style

➤ Damaged finish

➤ Reconfigured bathroom

➤ Desire for a higher-quality fixture

Many people love old cast-iron tubs and consider them worth keeping; others want to cast them out as a vestige of days gone by and good riddance. Damaged and scratched finishes that have seen better days aren't all that appealing to the discerning bather. And there's no way to dress up a cheap, bottom-of-the-line steel tub if you happen to have one of those.

Plumbing the Depths

When an automobile is factory painted, the body is dunked in rust-inhibiting primer and then is robotically sprayed with high-tech, coatings full of toxic solvents before being baked at high temperatures. Individual body shops can do a commendable painting job, but they cannot reproduce a factory job. The same is true with bathtub refinishers: They cannot duplicate the original process of bonding porcelain to steel or cast iron. Refinishing will always be a compromising job.

A major reason for replacing a tub, of course, is to do a complete remodeling job on an existing bathroom. These jobs are usually done to update and replace fixtures, and sometimes expand the bathroom's size by breaking down a wall and grabbing some space in an adjoining room or closet. Still, before tossing everything out, consider the alternatives.

Bathtub Liners and Refinishing

The manufacturers and installers of bathtub liners love them, but their competitors in the bathtub refinishing field, when they're not touting one finish over another, aren't so sure. A bathtub liner is an acrylic or fiberglass product that is fitted directly over your old tub along with acrylic or fiberglass walls that are installed over your existing tile. It's secured with double-stick foam tape and caulk. Unlike bathtub refinishing, a seamless bathtub liner is installed in less than a day without the odors from curing finishes and solvents.

Refinishers claim that the liners are easily stained and can trap water between themselves and the tub when water seeps in around the drain. The water becomes stagnant and, well, you can guess the rest. Even among themselves, refinishers don't agree on the best process. Some use epoxy finishes, while others prefer urethane enamels, which they claim will last longer. Despite competing claims, warranties are limited—maybe five years, at best.

Both alternatives are initially less expensive than replacing your tub, but the issue is the length of service. You must ask yourself whether you're throwing your money away on a short-lived solution. You might take extraordinary care in maintaining a refinished tub, for instance, but it can still turn dull or chip.

Plumbing Perils

It seems like just about everything is a source of lead these days. Testing has shown that porcelain-coated bathtubs are a source of leachable lead—that is, lead that dissolves in the tub water. Children then ingest the lead when they swallow bath water or mouth their toys after they've been in the water. If this is a concern to you, the tub and the bath water can be tested for lead.

Grab Your Sledgehammer

Used steel tubs are almost never worth saving. Contractors and plumbers often use sledgehammers to break them up into smaller pieces for easier hauling. Before playing demolition derby with your tub, you have a few chores to do first:

➤ Shut off the water and remove the faucets, tub spout, and diverter.

➤ Remove the overflow cover, trip lever, tripwaste, stopper, and waste outlet.

➤ Cut away a few inches of tile or other wall covering all around the tub ledge.

➤ Remove any nails securing the flange to the wall studs.

Pipe Dreams

If you have an old tub, it may have some resale value to an architectural salvage firm. If nothing else, make some inquiries and see if anyone will haul it from your bathroom if you get it loose. Keep in mind that there is always the issue of damage to your property while carrying the tub down the stairs or injury to the individuals carrying the tub.

Plumbing Perils

Be sure to wear eye and ear protection when you break up an old tub with a sledgehammer. The noise really reverberates in such a small room. Either disposable ear protectors or the muff style will do the job for you.

Now the tub is free of any plumbing hookups. You can either remove it in one piece by pulling it away from the wall or break it up with a sledgehammer. *Be sure to cover the tub with a piece of plastic or an old blanket first!* You don't want to be hit with flying pieces of debris, especially bits of porcelain. Cast-iron tubs are heavy and will take two people to get them loose (and maybe three to carry them out). If the floor tile was installed against the outer wall of the tub, you'll have to pry the tub up and over the lip of the tile to remove it. It also might be necessary to remove some of the floor tile to do so.

You might find after the tub is removed that you can't get it out the door! It's rarely worth tearing out a wall out and removing a door jamb to preserve a vintage tub, especially given the array of reproduction tubs that are available. Grab a sledgehammer and save yourself a lot of trouble.

Removing an Enclosure

A fiberglass or acrylic tub/shower enclosure won't be as cumbersome to remove as a steel or cast-iron tub (and you won't need a sledgehammer). Remove the plumbing encumbrances (spouts, drain, faucets, and so on). Cut away a couple of inches of the drywall along the perimeter, and remove any nails securing the enclosure to the wall studs. Use a utility knife to cut any caulk lines where the enclosure meets the floor.

The enclosure can be removed in one piece or cut into smaller sections with a saber saw or reciprocating saw. Cut the sides *away* from the pipes so that you don't accidentally puncture them. There's absolutely no salvage value to an old fiberglass or acrylic tub and shower unit, so don't worry about being careful when removing it.

The Aftermath

After removing any tub that has been attached to your walls—a claw-foot tub is an exception because it stands independently—you'll have to rebuild the sections that were cut away. It's likely with a tile wall that all the tile will have to be replaced unless you were extraordinarily careful during the removal and can find a new tile pattern that matches or compliments the existing tile. If your old tile was installed over plaster or drywall, you should remove all this material and install cement backer-board (see Chapter 5) for a sound foundation. Your other choice is to install some kind of tub surround or a complete, one-piece tub/shower unit.

As you can see, there's more to installing a tub than just the tub itself. It's likely that you'll end up repairing your floor as well.

Choosing a Tub

Basically two types of tub users exist in the world:

1. Those who use tubs only for showering
2. Those who really enjoy a good soak

The former group just wants a tub and doesn't need to test out different models at a plumbing supplier or home improvement store. Just about anything will do. The second group treats buying a tub like buying a mattress: It has to be just so, and they won't hesitate to jump in any tub on the display floor (remove your shoes first, please). These folks will check for these considerations:

➤ Room and comfort (whether it will hold one or two people, for instance)

➤ Design and looks

➤ Overall dimensions

Comfort is subjective, but dimensions are very objective. You'll want to know whether you can get it in your door and into the bathroom and whether your bathroom will need to be reconfigured to install the tub. It will have to be a real killer tub to compel a homeowner to rebuild a bathroom to accommodate it.

Installing Your New Tub

One of your biggest chores will be simply getting the tub into the bathroom without scratching the walls or woodwork on the way in. Measure the tub you intend to purchase first, and check the measurements against the size of your bathroom door opening. Tubs and shower dimensions are pretty standardized these days to fit through, well, standard door measurements, but check anyway. Remove the bathroom door from its hinges, and set it aside in another room so that it's out of the way.

Completely clean away any debris from your tub removal and/or demolition. All new bathtubs will come with installation instructions, too, but here are the normal guidelines:

Plumbing Perils

You don't want to secure your tub flange with any fasteners that protrude out too far, such as screws and washers, because they'll get in the way of the drywall or backerboard when they are installed over the flange. Instead, use roofing nails used because these nails have broad, flat heads.

➤ Install a level, horizontal two-by-four across the studs as a ledger to support the edge of the tub; place the stringer so that it will accommodate the highest point of the tub ledge or flange.

➤ Your greenboard or cementboard will be installed above or just over the flange, but it will not touch the ledge of the tub.

➤ The tub flange can be secured to the ledger with flat galvanized roofing nails. If there are no holes in the flange, nail over it, catching the edge only.

➤ Be sure that the tub is absolutely level (you don't want it shifting once it's full of water and a bather).

➤ Check that all connections, especially tub waste pipe, are securely installed.

➤ Use a half-inch branch line for the tub.

Kohler Memoirs Whirlpool and Shower Receptor.

Pipe Dreams

A low-flow shower head will save water without noticeably affecting the feel of the shower. Better yet, install shower heads with their own shut-off mechanisms so that you can better control the amount of water you use.

A single tub should slide uneventfully into the old tub's space. Be sure that it's level (use wood shims, if necessary, to make it so). Reconnect the plumbing (see Chapter 14, "Tubs and Showers"), and check for leaks.

After testing, you can repair the walls. If you're going to install a tub enclosure or new tile and a cement backerboard, you'll have to remove the faucet handles and shower head and arm again. This is a small inconvenience for being able to test your plumbing before you do your finish work.

Showers

A hot shower is one of the true pleasures of a modern plumbing system. You'd have to go back several generations to find housing that was built without bathroom showers. You might decide to install a separate showering area in your bathroom and use the tub solely for bathing and soaking. The easiest installation is a prefabricated acrylic or fiberglass shower unit that comes with three walls and a shower door. These units require three framed walls for support and installation. The walls should be insulated to deaden the sound from water hitting the enclosure. The alternative to a prefabricated unit is a custom shower stall that is typically fabricated with tile or glass construction and a separate shower pan.

The plumbing for a shower is fairly simple and calls for the following installations:

➤ A trap, a $1^1/_2$- or 2-inch drainpipe, and a vent pipe

➤ A half-inch hot and cold water branch line and a shower pipe

➤ A single or two-handled faucet and shower head

Plumbing Perils

Prefabricated acrylic showers and tub/shower units come with and without predrilled holes for the faucet(s) and shower arm. Be sure that these holes match your existing plumbing if you intend to reuse it. If you must drill your own holes, do so carefully and with a sharp drill bit. Take your time—too much pressure can crack the plastic.

Kohler Memoirs Shower Receptor.

You will be running three-quarter-inch distribution pipes through the walls until they reach the bathroom. Reducing tees (three-quarter inch to a half inch in size) will then

run the branch pipes to the shower, tub (if separate), lavatory, and toilet. Depending on the location of the soil stack, $1\frac{1}{2}$- or 2-inch drainpipe will slope directly into the stack or will connect to a series of similar size drainpipes that eventually empty into the soil stack. The vent pipe either will tie into the soil stack or will tie into another vent line that goes out through the roof.

Pipe Dreams

It's a good idea when installing two shower heads to make one of them a removable, handheld model. You then can use it to rinse the shower and tub when cleaning the bathroom.

Doubling Up

Showering together is an uneven experience at best when you have only one shower head. One of you is always cooling off while the other is warm. This particular shared experience calls for at least two shower heads.

Kohler Master Shower.

You can have as many shower heads as your system can handle. Multiple shower heads can be installed in any enclosed tub and come built-in with some acrylic tub/shower units. The shower heads themselves are relatively inexpensive, but each one must be connected to the main shower pipe because that's the one where the water temperature is mixed and diverted. To connect additional shower heads, follow these steps:

➤ Attach a tee to the shower pipe.

➤ Run a half-inch pipe to the second shower head.

➤ Attach a *drop ear elbow* to the end of the pipe, and attach your shower arm and shower head.

By adding a series of tees off each previous shower head, you can install a series of new shower heads.

Shower Doors and Curtains

Shower doors are a great convenience, whether they're sliding or hinged models. Kits come with a lower track that sits on the ledge of the tub and two sections that are screwed to opposite walls. The different sections are identified in the kit's instructions. The horizontal lower track must be cut to fit the length of the tub ledge and is held in place by the two vertical sections and silicone caulking. All shower doors are made from safety glass that cleans pretty easily, but the tracks do not. Cleaning the various surfaces and crevices that come with shower doors can be tedious.

A shower curtain, on the other hand, is simple to install and can be removed for cleaning or easy replacement. It doesn't look as elegant as a shower door, but simplicity has its place.

What's That Thingamajig?

A **drop ear elbow** is a threaded fitting with two ears that allow it to be nailed or screwed to a piece of bracing between studs to steady and secure a shower pipe or other water supply pipes. An ordinary threaded adapter isn't constructed to be secured this way.

Pipe Dreams

Whether your local code calls for it or not, be sure to install a pressure-balance shower valve. This valve senses a drop in either the hot or cold water pressure and adjusts the other water supply accordingly so that the temperature doesn't vary by more than two or three degrees. Always buy a brand-name faucet with a warranty.

Beyond Tubs and Showers

Fancy bathrooms are nothing new, but they are more common these days than a generation or two back. Tubs come in various shapes and sizes, and shower stalls can reproduce the feel of a rain storm or a waterfall. Some tub/shower units even come with built-in lights and stereos. All you need is a phone for ordering food and a keyboard for work and Internet access, and you'd never have to leave your bathroom.

Jetted tubs have been around for some years now, but think twice before installing one. As I mentioned in Chapter 9, "Choosing Your Fixtures and Faucets," some people find that they don't use jetted tubs as much as they thought they would after they've been installed.

Steam showers are one of the up-and-coming bathroom improvements that are arguably more useful and less cumbersome to install than a jetted pool.

Plumbing Perils

Shower rods that stay in place by spring tension have a bad habit of slipping and marking painted walls. Either put a plastic coaster under each end of the rod first or install a rod that's screwed to the walls.

Jetted Bathtubs

Often referred to by the brand name Jacuzzi, a jetted tub or whirlpool bath comes with its own framing recommendations, depending on the model. The tub must be supported by the floor and at the rim by the enclosure. The nailing flange is not meant as a tub support, but a means of sealing the tub from water seepage. Jetted tubs come with their own electrical requirements, including these:

➤ The installer should provide an access panel to the pump.

➤ A dedicated circuit should be run to the tub controls.

➤ The work should be done only under permit with a follow-up inspection.

If your jetted tub isn't wired correctly, it can void the warranty for the unit.

Plumbing Perils

Installing a jetted pool is more involved than installing a standard tub. If it's not set and sealed properly in its framing, it can be an expensive mess to redo. This happened to one electrician I know who spent thousands of dollars removing a jetted tub, repairing and tiling the walls, and reinstalling the tub because it wasn't done properly the first time. You should think about leaving this one to professionals.

A Private Steam Room

Steam shower is the more accurate term here, but most of us know what a steam room is even if we've never used one. In the movies, it's one place where spies and nefarious characters meet and discuss a nasty piece of business. Once a feature exclusively in spas or high-end hotels and a few homes, steam showers are increasingly being installed in residential settings. The easiest steam showers to install are prefabricated units or steam components that you can add to your existing shower.

Prefabricated acrylic units are terrific: no seams and no seals to worry about except the door. A big concern, obviously, is completely sealing the steam shower so that the water vapor doesn't deteriorate your walls or ceiling.

If you're thinking of custom building a steam shower instead of using a prefabricated unit, keep a few guidelines in mind:

➤ Install as long a bench as the space can accommodate.

➤ Place the steam outlet about a foot and a half off the floor.

➤ Slope the ceiling so that water runs down it rather than drips on you.

➤ Use cement backerboard or cement in wire mesh as a tile underlayment; cover with a coat of waterproofing thin-set.

➤ Thoroughly seal the tile grout with silicone sealer.

After using both prefabricated steam rooms and custom-built tile ones, I vote for prefabricated any day. They're completely sealed, easy to clean, and require almost no maintenance to the acrylic surface. A steam shower requires a hookup to a water supply pipe, a drain, and an electrical connection. Its plumbing requirements are similar to a shower stall.

The Least You Need to Know

➤ Both tub and shower replacements are big jobs and shouldn't be decided lightly; be sure that you're ready for the mess and expense before you start yanking out your old tub.

➤ Sometimes the most useful tool for removing a steel or cast-iron tub is a large sledgehammer.

➤ Tub replacement doesn't end with the tub; you'll also have to do wall repair and possibly floor repairs, too.

➤ Installing multiple shower heads is now the trend, and you should figure on installing at least two if you're doing a major shower installation.

➤ Think twice before installing a jetted tub, but take a closer look at steam showers, too.

Kitchen Concerns

Kitchens have evolved from strictly functional rooms for storing and preparing food to the center of the family universe. Part of this is due to our preoccupation with eating, but a behavioral anthropologist might argue that it's a return to the hearth from our cave days. That might be, but even the Flintstones didn't have commercial gas ranges, maple cabinets, and ice water dispensers.

Plumbing plays an important role in the kitchen. Few new homes are built without dishwashers, waste disposers, and double sinks. Many have ice and cold water dispensers built into their refrigerators, as well as instant hot water dispensers under the sinks. Some even have second, smaller sinks installed so that two people can work more independently. An existing kitchen might have usable plumbing, but you might want to replace the sink and appliances. Older kitchens are good candidates for new water pipes and drainpipes.

Kitchen remodels involve several trades, and you'll have to keep them in mind when planning your plumbing upgrades or changes. Cabinetmakers and the contractor doing the finish counter work need to know sink sizes and the location of faucets.

Your electrician must know where to install dedicated receptacles for the various appliances. Your dishwasher and sink colors will affect your color choice of other appliances and counter and backsplash materials.

Plumbing Perils

Installing a new kitchen requires that the work of various trades be scheduled carefully so that no one's work is held up. Plumbers and electricians usually get started early doing their rough-in work—that is, installing pipes and wires in the walls. When carpenters, cabinetmakers, and others are finished, a plumber will return to install the sink(s), dishwasher, and related appliances.

Pipe Dreams

A stainless steel sink is easy to care for and goes with any decor. More expensive models are made of a heavier-grade steel and are quieter when water hits the bowl or flatware is clattering around. Various cleaners for stainless steel will keep these sinks looking new for years.

Each of your kitchen fixtures should have its own shut-off valves, just as the dishwasher should. Dishwashers and disposers have very specific installation requirements that must follow both plumbing and electrical codes. Like any other fixture or plumbing appliance in your house, those in your kitchen will occasionally need repairs (sink clogs top the list). We'll cover those and new installations as well in this chapter.

Only the Kitchen Sink

The only plumbing fixture you really need in a kitchen is the sink. We can get along without disposers, dishwashers, and hot or cold water dispensers, but a good sink is a must. A kitchen sink can be a basic stainless steel affair with a simple double-handle faucet to a triple-bowl, color-coordinated, center-of-attention model with a high-end faucet, sprayer, and built-in soap dispenser. They'll both do the same thing: dispense water for washing, hold water when the drain hole is plugged, and carry waste water away.

Aside from the main sink, you may have, or may want to install, a second sink in another work space such as an island. This is a more involved installation than a typical sink that has easy access to the DWV system because it's located at an exterior wall. A second sink in an island will require a loop vent (consult your local code).

When you decide to replace an existing sink or install a new sink during a major kitchen remodel, note these two guidelines:

1. Be sure that your new sink will fit in your cabinets.

2. Your choice of sinks must match your choice of faucets.

Your cabinet installer will want to know the sink size so that the counter can be cut to properly accommodate it. *Don't buy a different sink without informing the cabinet contractor!* A larger hole can often be cut, but an existing hole can't be shrunk, and you may have to pay for a new counter. For installation of sinks and faucets, see Chapter 15, "Replacing Sinks and Faucets."

Dump and Grind

A waste disposer is a really fun gadget. Installed under your kitchen sink, a disposer's spinning blades or impellers turn organic food waste into a stream of finely ground particles mixed with water, essentially liquefying your leftovers. They then go down your drainpipe on their way to your local sewage treatment plant or septic tank. Proponents argue that disposers are environmentally friendly because they reduce food waste that, unless it's properly managed in a compost bin, will decompose in landfills, produce methane gas, and allow acid leachate to seep into the ground. Aside from that, old food sitting in the trash can really smells.

Opponents, on the other hand, would declare food disposers to be another evil appliance requiring fossil fuel-operated power plants to run them instead of homeowners properly recycling and composting their food scraps. This latter activity is hardly likely to occur with most urban dwellers, though, so I recommend the disposer over tossing your cantaloupe rinds into your trash cans during a summer hot spell.

You'll find two types of disposers:

1. Continuous feed
2. Batch feed

A continuous-feed model has a separate switch on the wall or inside the cabinet under the sink. The disposer will run as long as the switch is on. A batch-feed model comes with a special stopper which, when inserted and turned, activates the disposer. Once it's removed, the motor stops. Obviously, the continuous feed model is more convenient, but it could be argued that the batch feed is safer because there's no way that anything can accidentally drop inside while the disposer is running.

Plumbing Perils

Purchasing a disposer is no time to go the budget route. Buy the most powerful model available. They'll grind up anything, will have stronger motors and blades, and won't disappoint. There isn't enough difference in cost between low- and high-end models to justify buying a less expensive one.

Pipe Dreams

A disposer requires its own dedicated electrical circuit, which must be inspected. The electrical inspection can be scheduled independent of any plumbing inspection that your building code might require. If you know that you're going to be installing a disposer at some point in the future, go ahead and do the electrical work ahead of time so that it will be inspected and ready to go.

Ensuring That Your Disposer and Septic System Get Along

According to the manufacturer of In-Sink-Erator disposers, a septic system that can accommodate a dishwasher or a washing machine can also accommodate a disposer. After all, most of the waste is simply water. In-Sink-Erator has introduced the Septic Disposer specifically for septic systems. This model injects an enzyme treatment directly into the disposer's grind chamber every time the disposer is used. The enzymes assist the septic system in breaking down the ground-up food.

Disposer Rules

Nothing escapes the scrutiny of academia, and disposers are no exception. The following care and feeding of disposers was written in 1998 by Anne Field, Extension Specialist, Emeritus, from Michigan State University Extension Service (with references from the Maytag Corporation).

When a Disposer Is Indisposed

Quality disposers are very reliable appliances. Any disposer is subject to three common problems, though:

➤ The blades stop spinning because of an obstruction.

➤ The motor overheats and stops running.

➤ The drain becomes clogged.

Pipe Dreams

Before you have an old disposer repaired, compare the minimum repair cost with the price of replacement. If you have a really old unit, you might be better off installing a new one.

Plumbing the Depths

Water utilities really prefer that you allow your fats and grease to solidify in a disposable container and then toss them in the trash. You may be able to wash the fats out of your sewer line, but they can accumulate in the utility's lines and then must be cleaned out. The utility must deal with a system that serves thousands and thousands of households, so all that fat really adds up. It does make you wonder what the solid waste management people think, however.

Care and Feeding of Your Food Waste Disposer

Use a strong flow of cold water, and keep it running at least 30 seconds after the noise of grinding has stopped to flush all food particles through the drain line.

Always use cold water when operating the disposer to solidify fatty and greasy wastes so that they will be chopped up and flushed down the drain.

Hot water will not hurt the disposer, and you may safely run hot water from the sink through it. However, use cold water when you are operating the disposer.

If you wash dishes in a sink with a disposer, check to be sure that all small objects are removed from the sudsy water before you drain the sink.

If you have a continuous-feed disposer, move silverware and other small items away from the edge of the sink counter to avoid accidentally knocking them in while it is running.

Do put small bones through; they help to scour the sides of the grinding chamber.

Follow directions in the manual with your disposer as to what should not be put through it. Do not grind large bones or fibrous materials such as corn husks unless the manual says that you can. With fibrous foods (celery, chard, asparagus ends, and similar items) put through only a small amount at a time with a full flow of water. If the drain line is long and quite horizontal, fibrous foods or too much garbage at one time can clog the line.

Do not put uncooked fat off meat into the disposer, as it may clog. Do not pour liquid fats down the line; solidify them in an empty tin can in the refrigerator, and dispose in the trash.

Run the disposer each time you put food waste in it. This is particularly advisable in the less expensive models, which are more subject to corrosion from the acids formed by food waste left for a long time.

An unusual noise while the disposer is operating could mean that a foreign object is in the machine. Turn off the disposer immediately and retrieve the object.

With a continuous-feed model, use the cover as directed to protect yourself when grinding bones or fruit pits—small particles could possibly be ejected by the force of the disposer action. Avoid leaning over the disposer if you are feeding waste into it while it is running.

Never put your hand inside the disposer while it is running.

Grinding citrus fruit peelings or even half a lemon will freshen up the smell of your disposer.

A disposer can grind up all kinds of organic food waste, but it isn't always the tough meat bones that stop the blades from grinding. Fibrous waste such as corn husks or artichokes or even excessive amounts of celery can do the job. Of course, bones and steel flatware might do the job, too. To clear out the obstruction, follow these steps:

➤ Shut off the water and the disposer. It's not a bad idea to unplug the disposer, too, or turn off the circuit breaker if it's directly wired to its electrical circuit.

➤ Insert the Allen wrench that comes with the disposer into the hex-shaped center hole at the bottom of the disposer, and work it back and forth until it moves around freely in a circle.

➤ Shine a flashlight down the disposer, and remove the obstruction with a pair of kitchen tongs; if you must put your hands down the disposer, be sure that the power to the unit is turned off.

➤ If the unit is too tight to move with the wrench, remove any loose material from the disposer and, with the power off, stick a pry bar or the handle end of a plunger inside, forcing it against one of the grinding protrusions or impellers until the unit moves freely.

A good way to prevent flatware and other objects from falling down the disposer is to use the stopper that comes with the unit and covers the sink flange. If your disposer is running but sounds like it has a small bit of food that it's not quite expelling, toss a few ice cubes down to dislodge the food.

This Motor's Hot

A disposer is designed to shut off if the motor is being overloaded. This happens when an object jams the blades and the unit continues to run until it overheats. Once you've removed the obstruction, wait 10 minutes with the switch off (or the stopper removed from batch-feed models) and allow the motor to cool. Then push in the restart button on the bottom of the unit. If the disposer will not start after you turn on the switch, turn off the switch and check your electrical service panel for a tripped breaker or your fuse box for a burned-out fuse.

Bogged Down with Clogs

Having a disposer doesn't give any of us a license to stuff it with piles of food wastes and expect it to grind happily away. Wastes have to be fed gradually with plenty of water running. You'll know soon enough if your feeding habits are appropriate when ground-up food and water start coming back up and into the sink. Your clog might be in the discharge tube or the trap. Either way, you can often force it out with a plunger.

Oddly enough, if you toss too much ice down a disposer while it's running, the ice will turn to mush and clog it until it melts. Trust me on this one.

Disposer Hygiene

A smelly disposer is usually one with old food waste or built-up grease that hasn't flushed out. Commercial foaming cleaning products are available to help clear away these residues. Home remedies include these:

➤ Periodically fill the sink with hot, soapy water and pull out the stopper, flooding the disposer. The weight and volume of the water helps keep drains clear.

➤ Pour a cup or more of table salt into the disposer. Let it sit for a half hour, and then turn on the unit and flood it with cold water. This sometimes can eliminate odors.

➤ Fill the disposer with the contents of a small box of baking soda, followed by a bottle of vinegar. This will create a cleansing foam that cuts grease and fats. Flood the disposer with cold water to finish.

➤ With the disposer empty, turn it on while running hot water. Also give it a healthy squirt of liquid dishwashing soap that can cut through grease.

➤ If you've poured down an excessive amount of grease or fats, follow them up with a half-dozen ice cubes to help solidify them. Let the ice sit for an hour, and then run the disposer with cold water.

As you can see, you don't have to do much to keep your disposer in good working order.

Installing a New Disposer

Disposers come with detailed instructions that are pertinent to their specific models. A few key things to remember include these:

➤ The installation dimensions vary with each model and might not match your existing disposer if you're replacing an old unit.

Plumbing Perils

A smelly disposer could be an indicator of a problem with your DWV system, in which case the smell will be more like sewer gas than old food. If all your attempts to clean the disposer are unsuccessful, check that the water in the sink trap isn't being siphoned away.

Pipe Dreams

It's a good idea to write up a circuit map of your house, identifying all circuits and their breakers of fuses. You don't want to be guessing which one controls your disposer, dishwasher, or any other appliance when you need to work on one of them.

➤ Existing drain lines should be cleaned out with an auger before you install the new unit.

➤ The disposer should have its own trap, as should the other sink basin; most plumbers use what's called a disposal waste kit consisting of a P-trap and a baffle tee. This kit connects with a straight section of pipe from the disposer.

➤ Place a bucket or other container under your old disposer before removing it.

If you're installing your home's first disposer, be sure to install a dedicated electrical circuit first. A dedicated circuit can supply power to only one appliance.

Getting Dissed by Your Dishwasher

A dishwasher is kind of complex. It has a pump, a heater, hoses, a fan, and sprayers, and it isn't exactly consumer-friendly when it comes to repairs. Some problems are merely a matter of how you use your dishwasher. Leaks that appear to be caused by a deteriorated door gasket can result from bad dish-loading practices. If anything blocks the spray arm, it will shoot water directly on the door and will appear to be leaking. Whenever you load your dishwasher, check that nothing is blocking the way of any of your sprayers (you'll get cleaner dishes, too).

A stuck detergent dispenser cup can be another victim of bad rinsing. If the soap doesn't get rinsed out, it can obstruct the cup.

Plumbing the Depths

A built-in dishwasher requires a dedicated circuit, but it doesn't have to be ground–fault–circuit–interrupter (GFCI) protected. A portable dishwasher will use one of the existing counter receptacles. The electrical codes make a distinction between built-in appliances whose power source isn't very accessible, and portable or counter appliances whose receptacles can get you into trouble if you're not careful. Most local codes follow the National Electrical Code, but your local code is the one you must follow.

Repair or Replace?

Modern electric dishwashers are much more efficient than older models. They use less water and about one-third the energy of 20-year-old machines, according to

The Association of Home Appliance Manufacturers. Newer models also offer more washing and drying options and come with warranties. On top of that, new dishwashers are quieter, thanks to increased insulation. Consider the expense of repairing your current dishwasher with the cost—and cost savings—of a new model.

Energy Savings

It's tempting to run a dishwasher when it's only partially full, but just as much water and energy is used to wash one dish as to wash full racks. Washing full loads not only will cut your costs, but it also will mean less wear and tear on your dishwasher. By using the air-dry selection, or by opening the door of the dishwasher, you skip the heat cycle when drying the dishes. Of course, it can take forever for them to dry, but that's another matter.

Getting into Hot Water

Hot water dispensers, which heat water to approximately 190°, have become increasingly popular. The dispenser is installed under the kitchen sink and is connected to the cold water supply line. Cold water enters the tank, becomes heated, and is drawn out of a sink-mounted spout. The dispenser does not require a dedicated GFCI receptacle. According to the National Electrical Code, a hot water dispenser is viewed as a device that is fastened in place and whose receptacle is inaccessible to normal use. Your local electrical code, however, may require a GFCI, so be sure to check with your building department.

Hot water dispensers provide instant heated water for coffee, tea, and dehydrated soups. They're economical for heating up single cups instead of using a microwave oven or heating a kettle of water on the stove. And they're terrific for rinsing off really gunky items such as cheese graters.

Plumbing Perils

Dishwashers can be recalled just like automobiles. In 1999, the U.S. Consumer Product Safety Commission announced General Electric's voluntary recall of almost 3.1 million GE and Hotpoint dishwashers built between April 1983 and January 1989. A switch on the dishwashers presents a fire hazard, and consumers are warned to stop using the appliances. GE is offering a rebate toward the purchase of a new dishwasher; you can contact the company at 1-800-599-2929 or www.geappliancerecall.com.

Installing a Hot Water Dispenser

Like all plumbing appliances, hot water dispensers come with specific directions from each individual manufacturer. This installation can be done with a saddle valve, a handy device for tapping into a copper water supply line without having to use compression fittings or soldering the fitting. A saddle valve cannot be used on a flexible supply line. To install a saddle valve for a hot water dispenser, follow these steps:

➤ Shut off the water to the cold water supply pipe, and open the cold water tap at the sink to empty out the line.

➤ Position the valve on the cold water pipe so that the dispenser's feeder line can comfortably reach it and be tightened.

➤ Turn the valve's knob clockwise until the needle pierces the pipe and is securely seated.

➤ Open the cold water shutoff valve (or the main shutoff, if there is no individual shutoff), and check for leaks in the saddle valve.

Saddle valves connect an existing water supply pipe to a small branch line that feeds a particular appliance such as a hot water dispenser or an ice/cold water dispenser. After the dispenser has been installed—don't plug it in yet—open the saddle valve and check for leaks. Open the unit's tap to clear any air from the line, and check that the water is flowing. Then plug the dispenser in to the receptacle.

The simplest valve yet—
Saddle valve.

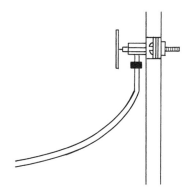

The Least You Need to Know

➤ Sink, faucet, and counter measurements all work together when installing a new kitchen sink.

➤ A food disposer has its limits; knowing them will give you fewer disposer problems.

➤ Dishwashers, disposers, and hot water dispensers all have special electrical requirements issued by your local building department.

➤ Overloading a dishwasher can result in improperly rinsed dishes and possibly leaks through the door.

➤ A hot water dispenser can be an economical alternative to heating hot water in a microwave oven or on the stove.

Part 5
Major Upgrades and Remodeling

Your plumbing is one of your home's major mechanical systems. Just about any major remodeling job will include some plumbing upgrading (or a lot of upgrading). Much of this is behind-the-scenes work that you never see, but you enjoy its results every time you step into your new bathroom or put a glass under your ice dispenser in the kitchen. Aside from improving your enjoyment of your home, an updated plumbing system also adds to its value.

Despite the advantages of upgrading, new plumbing isn't as instantly gratifying as, say, refinished floors that you see every day. And it means tearing into your walls and floors, which is hardly an enticement. This part introduces you to the methods—without the madness—of big plumbing projects. These are major budget items, so you'll want to go into them with your eyes open and informed.

We'll also discuss your outdoor plumbing, an area that includes getting water to your garden hoses and installing complete sprinkler systems. Be prepared for some digging, though—those pipes have to be well buried to prevent them from freezing. Once they're in, you'll forget about all the work when your sprinklers automatically water your yard. There's a lot to be said for no-brainer yard irrigation.

The Big Picture

Small plumbing repairs or improvements can be done without much effect on the surrounding scenery. A new kitchen sink with the same dimensions as the old sink won't require that you alter the counter, for example. But what if you want a new sink that's much larger and your dishwasher's location prevents you from installing it? Suddenly your plumbing project becomes a larger remodeling project.

You might be planning a major project, such as remodeling a kitchen or adding a new bathroom. Plumbing will be only one of several crafts needed to complete these jobs. Planning will be more critical as you figure in the different trades, materials, permits, and inspections required for your undertaking. Budget considerations will take on a new meaning. After all, the cost of a bathroom fixture is one thing, while the cost of a new bathroom moves the decimal point over a notch or two on the final dollar amount.

Your skills might lie in another craft besides plumbing, or you might be hiring out the entire job. You'll need to know the specifics of each phase of the work and the order of events (which trade comes first, who finishes last). This chapter introduces you to the remodeling tasks sometimes associated with plumbing, everything from resurfacing an existing counter to adding an additional bathroom. Kitchen and bathroom remodeling are typically the most expensive changes done to a home, but at least you'll have some idea what you're getting yourself into.

Starting Small

Sometimes our existing homes are pretty close to what we want, but something just isn't right (or sometimes a lot of things aren't right). It might be the carpet or the wallpaper in the upstairs hallway. The 1930s bathroom tile might appeal to your artistic sense, but the hanging wall sink with the rust stains and separate hot and cold water spigots might be a little too quaint. Replacing the sink with a contemporary model brings up several questions:

➤ Can you match the color of your existing fixtures?

➤ Do you install another wall-mounted sink or something else, such as a vanity?

➤ If you install a pedestal sink, how do you cover up the holes in the tile from the old wall-mounting brackets?

Color matching isn't much of a problem if your fixtures are white. Sure, there are different shades of white, but you should be able to live with minor variations. A white sink and a sea-green toilet and tub is a different story.

A wall-mounted sink is secured to the wall with mounting brackets; the brackets are screwed through the tile and into the wall. Removing these will leave holes. No matter how close you can match the color when you patch up the holes, you'll always see the patch. Installing a vanity will solve that issue, but what if you don't want a vanity and had your heart set on a standing pedestal sink? How do you cover up those holes so that they're less visible?

Pipe Dreams

Reproduction fixtures and faucets are available to match all kinds of existing styles, from claw-foot bathtubs to art deco faucets. Plumbing showrooms will have plenty of catalogs to look through if you don't see what you want on display.

Your plumber, if you're hiring the job out, won't patch your tile for you, so consider alternatives. Fill the holes with grout or mortar, and glue a plain brass or polished metal button over the hole. A stroll through your hardware store or a sewing center will present all kinds of alternatives.

Counter Culture

Both kitchen counters and bathroom vanity tops are often covered with plastic laminate, a sanitary, easy-to-clean material available in a huge selection of colors and designs. This is a terrific surface finish, but scratches will show, especially on laminate with a glossy finish. Old plastic laminate can look even more dated next to a shiny new sink, but a solution is at hand.

New laminate usually can be applied over existing laminate if the old material is properly prepared. It must be sufficiently sanded and "roughed up" to allow the new material to bond. *This is not a job for an amateur!* Laminate work is very exacting: The material must be cut and fitted with the fewest seams possible and glued into place with a toxic-smelling, fast-drying adhesive. Trimming and finish work is done with a router and a hand file. This is one task to hire out.

If you don't want to install laminate again, other materials, including tile and wood, can usually be installed over existing laminate, although both will raise the height of the counter.

Plumbing the Depths

Our last home was a 1924 bungalow with much of the original kitchen. There was little room to expand, so rather than spend thousands of dollars updating a limited space, we added a bit to the existing cabinets, installed a food disposer, and had one cabinet gutted to make room for a dishwasher. Then we installed a new wider counter, thus covering the dishwasher. These were simple, inexpensive procedures that significantly updated the kitchen.

New Appliances, Old Kitchen

You'd be surprised how adaptable an existing kitchen is to the installation of new appliances. Even very old kitchens built before the existence of dishwashers can be tweaked and stretched to accommodate disposers, extra sinks, and dishwashers, of course. A talented cabinetmaker can remove existing drawers or a cabinet and extend a counter to cover a newly installed dishwasher.

When Electricity and Water Mix

The plumbing and electrical systems are two distinct systems in your home, but they often intertwine. For example, a dishwasher and a hot water

What's That Thingamajig?

A **dedicated electrical circuit** is one that supplies power only to a specific appliance or receptacles. This is done so that the circuit doesn't get overloaded with demand for electrical current and shut down by tripping a circuit breaker or blowing out a fuse.

dispenser both require dedicated electrical receptacles. To meet the current electrical codes, bathrooms and kitchens should be equipped with GFCI receptacles. Simply because you're updating your plumbing does not mean that you're required to update your wiring, but you might not have much choice in some instances. A dishwasher, for example, requires a dedicated circuit, and so does a disposer.

If you have a really old fuse-based electrical system, it might not be capable of accommodating any additional electrical load, such as a dishwasher or laundry appliances (if you don't already have a washer and dryer). If you have any questions regarding your electrical system, have it evaluated by a qualified electrician before you order a new dishwasher.

Serious Remodeling

Houses of all ages are subject to remodeling as new owners settle in and find that they want another bathroom or an expanded kitchen. These are obviously larger jobs than simply adding an appliance or changing out a fixture. Adding a new bathroom, for instance, will require the following:

➤ A set of plans

➤ Permits and inspections for the framing, electrical, plumbing, and sometimes the insulation

➤ Framing, insulation, heating, and ventilation

➤ A roof, if the room is being added on beyond an existing outside wall (same for exterior siding and painting)

➤ Plumbing and fixtures

➤ A GFCI receptacle and lights

➤ Installation of a floor covering

➤ Drywall and painting

➤ A door, possibly a window, and storage

This is a big project, almost a house in miniature if you think about all the crafts involved. Converting an existing room to a bathroom saves you some of these steps, but you'll still have to cut into existing walls and the floor and patch up afterward.

Pipe Dreams

Install a substantial fan in any new or remodeled bathroom (a fan is required in new bathrooms). You want the moisture out quickly rather than having the fan drone on and on trying to clear the room. Buy a fan that removes at least 100 cubic feet per minute (CFM) or larger, even though it goes against common building practice to install fans with a minimum CFM power.

Plans? What Plans?

Some building departments require a drafted set of plans for a bathroom, and some need only a general outline. Discuss your addition or remodel with your building department first, and find out what is needed before you can get your permit. A set of plans will accomplish several tasks:

➤ They allow any bidding contractors to offer a clear and precise estimate for the work.

➤ They show any problems with the design, such as fixtures in the way of swinging doors.

➤ They provide an outline of materials needed to complete the job (you'll need a separate, specific materials list, however).

It's a *lot* better to discover any design or space problems by reviewing the plans than running into them after you've finished framing. Plans also allow a contractor to suggest alternatives or changes that you might not have considered, such as the locations of pipes.

Plumbing Perils

Be sure to look at fixtures and materials well in advance of your remodeling. Browse home improvement stores and plumbing showrooms. You'll get an accurate idea of costs and can set a realistic budget instead of being surprised close to the start of the project.)

A Raft of Crafts

Adding a bathroom or extensively overhauling an existing one will require the work of several different trades and the use of a variety of materials. Trades include these:

➤ Carpentry
➤ Plumbing
➤ Electrical
➤ Drywall installation
➤ Floor/counter installation
➤ Roofing
➤ Painting

Materials include these:

➤ Framing lumber
➤ Finish trim
➤ Insulation
➤ Subflooring or underlayment
➤ Drywall
➤ Plumbing fixtures and pipe
➤ Electrical wiring, devices, and fixtures (receptacles, switches, lights, heater, and a ventilation fan)

➤ Tile, vinyl, or wood flooring

➤ Tub and shower surrounds (tile, stone, laminate, or acrylic)

➤ Shower door

➤ Vanity

➤ Paint, wallpaper

Your list might be even more extensive. You can always add towel warmers, a leaded glass window, and even a heated floor. The point is that there's a lot of material to install, and its installation must be coordinated. You don't want to be painting walls before the shower is in place. The usual order of work goes something like this:

An existing bathroom is stripped of any fixtures, trim, cabinets, and wall and floor coverings that are to be replaced; a new bathroom will be framed after removing any necessary existing wall(s).

All mechanical (plumbing, electrical, heating, and ventilation) rough-in work is done.

Rough-in and framing inspections are done, insulation is added (this might be inspected with framing or could require a separate inspection).

After rough-in work is inspected and passed, the drywall and subflooring can be installed. The tub and shower are normally installed before the drywall.

Drywall is taped and finished, electrical fixtures and devices are installed, and tile work starts.

The finished floor may or may not be installed before the vanity or built-in cabinets and lavatory. The toilet is installed after the finished floor.

Finish trim is installed (some may go on after walls are painted or papered).

Walls and ceiling are painted or otherwise finished.

Electrical cover plates and trim for lights are installed.

Clean-up takes care of any mess.

The process is similar for kitchens except that more appliances and cabinets are being installed. Any windows will be installed after the framing is completed. If you're hiring out all or any of this work, each trade or craft will have to be scheduled and coordinated. A contractor's time is valuable (so is yours), and you don't want someone showing up when the job site isn't ready. A tile contractor, for example, can't afford to schedule your job and then lose a day because your walls aren't completed. One solution is to hire a general contractor to do the job.

Generals and Subs

A general contractor is licensed to do major re-modeling and building jobs, but he isn't necessarily licensed to perform a specific trade such as electrical work. The rules vary from one state to another, however, and what's true in Portland, Oregon, might not be true in Atlantic City, New Jersey (I'll give you 4 to 1 odds against it). This work is usually done by a subcontractor—an electrician, in this case—who specializes in one trade. However, many general contractors will perform at least some tasks usually done by specialty contractors, providing that there are no legal prohibitions from doing so. For instance, although all contractors need to be licensed, only a few trades are obligated to pass tests to practice their craft. Plumbers and electricians usually have to pass tests, but a tile installer typically does not. A general contractor with a talent for installing plastic laminate or roofing might choose to do that work instead of hiring it out separately.

When you hire a general contractor, you're paying for building and supervisory skills. The price you are quoted is based on time, materials, overhead, and profit. The time factor involves hiring and paying subcontractors, ordering and picking up materials, meeting with you and your architect (if one is involved), and office time. Terms of your contract will vary from cost based on time and materials (you pay per hour until the job is done, plus the cost of materials) to a fixed bid (plus an allowance for any changes to the job).

Plumbing Perils

It's a good idea to expose your kids to remodeling work, but they shouldn't play around a work site unsupervised. There are too many tools, protrusions, and holes in floors that can cause injuries.

Pipe Dreams

Some contractors specialize in bathrooms and kitchens and often have more experience in these areas than other general contractors. Look at some of their recent work and see whether they offer a creative edge you'd like to see in your own bathroom or kitchen work.

Floors and More

Painted wood floors typically were used in common residential bathrooms and kitchens until linoleum came into use in the latter part of the nineteenth century. Wood and water aren't the greatest combination of materials, and more than a few wood floors rotted out in bathrooms from leaky tubs. Linoleum—and, later, vinyl—floors protected the wood underneath and was a huge improvement in sanitation. Oddly, finished wood floors have been finding their ways back into kitchens and bathrooms.

You have an array of flooring materials to choose from for your bathrooms and kitchens, from the practical to the ridiculous:

➤ Vinyl

➤ Tile

➤ Marble

➤ Stone (granite, slate, and so on)

➤ Wood

➤ Carpet

Each material comes with its own unique properties and installation procedures. Vinyl, either in the form of individual tiles or continuous sheets, is the easiest to install and maintain. It comes in a sizable selection of colors and patters to choose among, it's easy on the feet, and it survives spills and dropped baby food. Some people see vinyl as being either old-fashioned or lacking in pizzazz, though, which is unfortunate because this is a great floor covering for bathrooms and kitchens.

Plumbing Perils

With vinyl flooring, you get what you pay for. Avoid the low-end, thin material, and buy a thick, high-quality product that will give you years of service and enjoyment.

Vinyl flooring is installed over a flat, level subfloor or underlayment of plywood or some form of particle board, which is a manufactured wood product. An adhesive is spread on the subfloor, and the vinyl is laid over it and rolled out to force out any air pockets. Vinyl tile is laid down piece by piece, allowing for different patterns in the floor. Installing vinyl flooring is another nuisance job of cutting, fitting, and precise laying; that's why there are vinyl floor contractors in the world.

Tile, Marble, and Stone

If you want a permanent material under your feet, albeit one that's cold to the touch and fairly unforgiving, then tile is the most affordable choice (besides plain concrete, anyway). Marble, either in sheets or tile, and the varieties of stone are costlier than tile, but they also provide a very elegant finish. Choose carefully! Once you've installed this material, you don't want to be yanking it out because you don't like the color or the pattern.

The best way to install these finish floor materials is over an underlayment of mortar (mud set), professionally installed over wire mesh and then leveled. It should last as long as the house exists. Cement backerboard, which is basically a four-by-eight sheet composed of cement and fiberglass, is an alternative to mortar.

Plumbing the Depths

Adding tile, marble, or stone to an existing floor can raise it an inch or more, but you can work around it easily enough. Thresholds can be installed to make an easy transition to adjoining rooms and hallways. Doors can be trimmed down to clear the higher floor. To be sure that the floor isn't going to be objectionably high, place a sample piece on the existing floor, walk on it, and see if you can live with it visually.

Dumb Floors: Wood and Carpet

We bought a new house in 1994. It had—and still has—a carpeted master bathroom, a quick and easy way for a contractor to finish off this room. It's soft under our feet but is an extremely poor choice for a bathroom floor. Carpet doesn't dry out quickly once it's wet, and this means it's an ideal spot for spores to call home. If you have kids, carpet becomes a spore metropolis. Why install something you can't really keep clean and that can affect your subflooring if it gets too wet? Install vinyl instead, and use washable throw rugs if you want something cushy under your feet.

As far as carpet in a kitchen goes, tear it out if you have it, and don't install it if you're thinking about it.

Wood floors are another poor choice for bathrooms and kitchens. They will show every single scratch and ding in a kitchen and can collect water if there's any separation between any of the boards. Refinishing a wood floor is a major job that's often dusty (new dust-free sanders are available, but refinishing contractors charge more for their use than they do for traditional sanders), and you, your kids, or your pets will start scratching them almost immediately simply from day-to-day use. You've probably guessed by now that I'm not a fan of wood floors in either kitchens or bathrooms, and I can even speak from personal experience: Our house came with these, too.

Walls and Finishes

Years ago, before texturing machines were invented, plaster walls were built with a smooth finish, especially if they were going to be painted. Drywall requires very exact finishing to produce a smooth wall whose seams aren't noticeable under paint. Texture covers a number of imperfections in the *taping* and *mudding* of drywall, which means that the job gets done faster.

If you want wallpaper in your bathroom or kitchen, it's best to hang it over a smooth wall (some papers might work okay on a lightly textured wall). New drywall should be primed before hanging wallpaper. Bathrooms are moist places. You want a wall finish that you can clean, which limits your selection of wallpaper. A smooth vinyl will work, while grass cloth won't.

Moisture and ease of cleaning are two reasons you don't paint a bathroom with flat paint whose dull finish does not clean up well. Most new bathrooms and kitchens are painted in satin finish, which has just a hint of a sheen and is easily wiped clean.

What's That Thingamajig?

Mudding is the application of joint compound to fill the joints and nail holes on newly installed drywall. **Taping** is the application of a strip of either paper or fiberglass material into a bed of mud at the seams to fill and reinforce the mud. Additional layers of mud are spread over the embedded tape.

Plumbing Perils

Never drape any towels or clothing over built-in heaters, especially a resistance heater. Fire departments all over the country have gruesome statistics on fires caused by these very actions.

Some Like It Hot

A kitchen is a sizable enough addition or remodel that it probably will be heated by your furnace, although supplementary heat might be added in a breakfast nook in the form of a small electric heater such as a baseboard. The same is true with a new or added bathroom, except that supplemental heating is worth considering (nudity can be a chilling experience).

The most common bathroom heaters include these:

➤ Baseboard heating

➤ An in-wall resistance heater

➤ An overhead heat lamp

Built-in heaters will require either a 240-volt or a 120-volt electrical circuit. This will be a dedicated circuit, although other heaters can run off a large enough circuit. A resistance heater has a small built-in fan that blows room air over a heating element. A baseboard unit contains an element that warms up as an electric current passes through it. The element's metal plates or fins direct the heat outward into the room. An overhead heat lamp is the least effective. Hot air rises, and with a heat lamp and its radiant heat, it starts out at the ceiling.

Putting It All Together

As you can see, plumbing is only part of the equation when it comes to adding on a bathroom or a kitchen or remodeling existing ones. Plumbing is a big part, certainly, but it must be coordinated with other tasks and crafts. This coordination is your job or that of your general contractor. Follow a logical construction order, and you shouldn't run into any problems.

The Least You Need to Know

➤ Upgrading parts of your plumbing system will often involve other house upgrades as well.

➤ Electrical upgrades go hand in hand with plumbing improvements, especially if a new appliance such as a dishwasher requires its own electrical circuit.

➤ Major remodeling projects are best planned out first so that you understand all the steps that are involved and can get a firm idea of costs.

➤ Coordinating the various jobs and trades on a remodel or building addition is the job of a general contractor, unless you decide to take on this task yourself.

➤ Choose your finish materials—flooring, shower and tub surrounds, counters, and wall finishes—carefully; each comes with its own installation and maintenance issues.

Chapter 19

The Basics: All New Pipes

In This Chapter

➤ Know thy plumbing code

➤ Pipes to bring in your clean water supply

➤ DWV pipes to get rid of waste water

➤ Pipe size counts

The ultimate in plumbing jobs is a complete repipe or replacement of all water supply pipes in your home. This doesn't have to be done solely as an act of desperation because your old galvanized pipes are springing new leaks every day. A general upgrade or a desire to avoid future problems with your present system is justification enough. Of course, if you have a suspect plumbing system, such as PB pipe, you would be more than warranted in replacing your pipes.

We discussed the different kinds of pipe in Chapter 8, "Pipes: Joining and Fitting." You may have several types to choose among depending on your local code and water conditions. Each will have specific installation procedures, but all of them will have something in common: They'll need to access the insides of your walls, the ground floor in most cases, and ceilings. This means cutting holes and later repairing those same holes. Plumbers will cut but not patch unless they include that in their bid, in which case they usually hire another contractor to do the work.

A repipe is a big job and doesn't lend itself to a do-it-yourselfer who can tackle it only part-time. Assuming that your new pipes can use the same holes in your house framing—joist, sill plates, and studs—you'll have to remove your old pipes before installing the new ones. If you're going to be drilling new access holes, then you could conceivably install a new system and keep your old one in place until you're ready to make the final connections at fixtures and faucets. At that point, you can gut out the old pipe. Either way, it's a time-consuming job. Consider whether it's better use of your time to hire it out and perhaps put in overtime at work (or moonlight), or whether you should do the job yourself.

This chapter goes over the basics of repiping.

Before You Start, Ask Yourself This ...

The effective life of a plumbing system can be measured in a couple ways. Ask yourself, does it still function sufficiently to satisfy its users? It doesn't have to work perfectly; it just has to work. If you don't mind a drain that's less than perfect or a galvanized pipe that's slowly clogging, then your system is suitable for your purposes. Another way to measure a system is to ask whether it can be better, whether it is working at its optimum performance, and whether you can trust it to continue and not develop a major problem when you least want it to. The answers to these questions might lead you to consider replacement.

The Code Comes to Dinner

A repipe certainly will introduce you to many of the intricacies of your local plumbing code. Any section of pipe that you aren't replacing—an existing iron soil pipe, for instance—most likely also will have to be brought up to code compliance. Galvanized pipe is out, so you'll have to choose among the following (depending on your code):

➤ Copper

➤ Chlorinated polyvinyl chloride (CPVC)

➤ Cross-linked polyethylene (PEX)

➤ Polyvinyl chloride (PVC)

The advantages of copper are well known: It's readily acceptable by every code, it's long-lasting, and it connects well to valves. Its downfall comes in areas where water is too acidic. Small pits start developing, and pits lead to leaks as the acid in the water eats away at the copper. Check with your local building department and water utility for a recommendation.

For the most part, you can't go wrong with an all-copper repipe. You'll use all three grades in your work:

➤ M, or thin wall pipe, for inside piping

➤ L, a thicker grade, for underground piping to sillcocks

➤ K, the thickest grade, for connecting the water meter to the water main

Plumbing the Depths

Don't let the additional cost of copper over plastic put you off. You will not install a new plumbing system again for as long as you own the house, so you might as well put in one that you know will last. Real estate agents regularly report that copper plumbing adds to the resale value of a home.

CPVC, an interior plastic pipe, should be considered if you have acidic water because it stands up to corrosion. PVC, on the other hand, is primarily used in irrigation systems and drain lines. PEX, a flexible pipe, is a superior successor to polybutelene (PB) pipe, but check that it's acceptable by your local code.

Pipes and More Pipes

You have four categories of pipe in your home:

➤ The main water supply pipe or service line

➤ Distribution pipes

➤ Branch pipes

➤ Supply tubes

It all starts with your service line. Its size is a big determinant of how easily you'll be able to supply water to your fixtures. The simplest advice? Install a 1-inch line and don't bother with a three-quarter-inch line, even if it meets the minimum code requirements.

Distribution pipes take up where your service line stops. These pipes are installed throughout your house to feed water to branch pipes. The size of your distribution pipes is determined by the number of fixtures and appliances they will be servicing, as well as the distance your pipes must cover from your curb stop to the fixture farthest away (a third-floor bathtub, for instance). Each fixture and

Plumbing Perils

If you want to switch from galvanized pipe to plastic pipe, you'll have to change the grounding for your electrical system if it is currently grounded to your existing pipes. Grounding to the pipe is considered a secondary grounding with newer systems, but it may be the primary one for yours. Have your system checked by an electrician if you have any questions. An electrical system cannot be grounded to plastic pipe.

appliance is measured in fixture units, and each different size of pipe can adequately provide water to a limited number of these units over a given length of pipe.

One-inch distribution pipes coming off a 1-inch service line will take care of just about any residential plumbing needs. For many homes, this can be overkill, and you can get by with three-quarter-inch distribution pipes. Besides, a 1-inch pipe holds more water in it than a three-quarter inch pipe. This means you'll have to empty it of more cool water first before you start drawing off any hot water from your water heater and this means a longer wait.

Plumbing the Depths

You can't determine the water pressure as it's supplied by your water utility; the utility controls it. You can affect it with your choice of pipe size, however. Water pressure and the rate at which water flows are very much related. It takes more pressure to push a given amount of water through a half-inch pipe than through a 1-inch pipe. The distance the water must travel is also a factor in the water flow you'll get when you open a faucet. Smaller pipes will react more adversely to a lower water pressure than larger pipes.

Branch Lines

Consider the branch lines to be the side streets coming off your main boulevard distribution pipes. Branch lines will be smaller diameter pipes because they're feeding only individual fixtures or appliances. The distribution pipes, on the other hand, must supply water to all the branch lines, so they are necessarily larger.

Most fixtures and appliances are easily supplied with half-inch branch lines.

Supply Tubes

As you get closer to actually hooking up a pipe with a fixture, it becomes more labor-intensive to connect the pipes. This is where flexible supply tubes come in handy. These tubes are either three-eighths or a half inch in diameter (three-eighths is more common) and are one of three types:

➤ Braided mesh
➤ Chromed copper
➤ CPVC plastic

Supply tubes attach to faucets, toilets, appliances, and shutoff valves with compression fittings. Flexible copper is also used and is relatively easy to work with, but not as easy as the braided mesh or plastic tubes.

Planning the Layout

There are all kinds of ways to layout your new system—some are just more tedious and expensive than others. Experienced plumbers take the shortest routes possible with the least amount of disruption in the form of torn-up walls and floors. On the other hand, because a plumber wants to save time, too, he or she may route your pipes in a way that's more convenient for installation purposes and with less thought toward wall and floor repairs.

Basically, from the point your distribution pipes take off from the service line, you will be running your pipe in a series of right angles. These right angles are formed with the help of a series of the following fittings:

➤ Unions

➤ Reducers

➤ Tees or T-fittings

➤ Elbows

Unions join two lengths of pipe for a continuous run. Reducers, and reducing tees and elbows, allow for the transition from one size of distribution pipe to a smaller-size branch pipe. Elbows connect two pipes at a corner, and tees allow for a break in either a horizontal or a vertical pipe to accommodate the insertion of a pipe going in the opposite direction while also continuing the original direction of the water.

Pipe layouts should be done on paper or computer first so that you can configure the most efficient design. In the course of writing up your layout, you'll have to allow for various obstacles in your house, such as these:

➤ Electrical wiring

➤ Furnace ducting

➤ Fire blocks inside walls

➤ Walls and floors

Plumbing Perils

Remember that pipes and drain lines make noise when they're in use. You don't want a toilet waste line running across your living room ceiling if you can avoid it. Plan your installations with this in mind as much as the convenience of the location.

What's That Thingamajig?

A **fire block** is a piece of two-by-four installed horizontally or angled between two wall studs. Fire blocks were installed to slow the spread of fire inside walls. You tend to find them in older homes.

Pipes and wires don't mix and cannot be installed in the same holes in joist or wall studs. You can't run either of them through furnace ducts either. Walls and floors have to be opened up, which means cutting through plaster or drywall. Doing a careful layout ahead of time will minimize your work, although it can't foresee every impediment, such as *fire blocks*.

Installing Water Pipes

Doing a total repipe on your own can be daunting and time consuming. If you can do the work only part-time and are learning as you go, how do you get the job done and convince your family to live without indoor plumbing for weeks on end? Telling your kids that it's National Back to Nature Month isn't going to cut it—and besides, they'd get on the Internet and figure out soon enough that there's no such event.

You can do the job in segments and cut off water to only one set of fixtures at a time. Here's how:

➤ Install a new service line.

➤ Remove the hot and cold water pipes from your first set of fixtures (say, a first-floor bathroom).

➤ Install the first length of distribution pipes to some point beyond this same set of fixtures. Install tees on the new sections of hot and cold pipes from which you will run branch lines to the first fixtures. (If you run the pipe up to the next set of fixtures, simply install another tee and tie it into the existing pipe.)

➤ Tie the new distribution pipe to the existing one with appropriate fittings (copper to galvanized, copper to copper, and so on).

➤ Install a shutoff valve off the tee so that you can continue working on the branch line without having to keep the rest of the water shut off.

➤ Run your branch line pipes and test for leaks.

This approach has several advantages:

➤ You can take your time establishing your branch lines while still keeping the water running in the rest of your house.

➤ You'll have plenty of shutoffs in the event of future repairs.

➤ You can take the work as far as you want and can call a plumber to complete the job if you run out of time.

Pipe Dreams

Whenever possible, plan your pipe routes so that you can reuse existing holes and notches cut into joists and studs. You can minimize the extent that you'll have to tear into walls if you can push your new pipe straight up through the plates at the base of the wall and connect to your fixture.

This approach means an extra fitting or two if you have to tie your new distribution pipe into your old one at a point short of a tee or elbow, but it's a small price to pay to keep your water on. Besides, you'll get more soldering practice if you're using copper or plastic pipe.

Getting Through the Framing

Pipes, like wires, have to pass through your home's framing. You can run your pipe one of two ways:

1. By drilling holes
2. By notching

Either approach has some limitations. Drilled holes must be at least five-eighths of an inch from the edge of a stud and 2 inches from the edge of a joist. This is to help prevent damage to the pipe when drywall is nailed or screwed on. Metal nailing plates (protector plates) should be nailed onto the edge of any joist or stud where a pipe passes through. These plates add further protection if a nail or screw is inserted near the hole or notch.

A notch is simply a small square of wood that is sawn and chiseled out of the edge of a joist or stud as an alternative to drilling a hole. You cannot drill as big a hole or chop out any size notch that you choose. You're limited by the size of the framing member. A load-bearing two-by-four stud, for example, can accommodate a hole diameter of one and seven-sixteenths inch or deep notch of seven-eighths of an inch. A two-by-ten joist can take a hole diameter of three and one-sixteenth inch or a deep notch of one and seven-eighths inch.

To cut a notch, follow these steps:

➤ Mark off the measurements with a pencil.

➤ Saw the vertical legs with a hand saw or an electric saw.

➤ Take a sharp chisel and cut out the horizontal section; you'll be chiseling along the grain of the wood, and your notch should come out easily.

In some cases, you'll be running more than one pipe through a joist or a stud. The pipes should be installed one over the other, not side by side. Give yourself enough distance between them that you can easily install them as well as any fittings.

Pipe Dreams

The bigger the drill, the faster you can drill out your holes for those pipes that cannot use existing holes or notches. If possible, go ahead and mark every location on a joist or plate that you intend to drill, and rent a 7.5-amp commercial drill. This way you can drill everything at once and incur only one rental charge.

Soldering Notice

Essentially, all soldering and flux used today must be lead-free. The limit is actually .2 of 1 percent, which is just about the same thing as lead-free. Do your soldering in a well-ventilated area.

Just a Few Rules

When you repipe your house, you will have to follow code rules regarding these elements:

➤ Sewer pipe installation

➤ Underground pipe

➤ Installation of shutoffs

➤ Pipe supports

➤ Slope requirements for drain pipe

➤ Venting

You can't go wrong if you do more than the code requires, but you must do the minimum requirements as directed by your local plumbing code. A national code may guide, but local code decides.

Plumbing Perils

If you do have to replace your sewer line and have been plagued by tree roots in the past, now is the time to deal with them. When digging the trench, cut away the roots. Consult with your local nursery for the best approach that will still safeguard the tree(s).

New Sewer Lines

You might find yourself laying a new sewer pipe or replacing an existing one. New pipe can be plastic, ductile iron, or vitrified clay with neoprene gaskets. Precast concrete and vitrified clay pipe were once popular choices, but heavy gauge plastic is a common choice today and with good reason: It's lightweight like its water supply pipe counterparts, it's easy to install, and it's long-lasting. Sewer pipe must be sloped downhill (sewage—and water, for that matter—can't run uphill). The grade varies depending on the diameter of the pipe, but you can figure a slope of at least a quarter-inch per foot of pipe. The pipe should be bedded in gravel and thoroughly supported (any low spots can allow waste water to collect). Your sewer line also must have an accessible cleanout.

Going Underground

Type K copper is a frequent choice to connect the water meter with the water main, but plastic is being used increasingly where the code allows it. Code requirements vary regarding main water supply lines, but some general guidelines follow:

➤ The supply pipe must be buried 1 foot below the frost line and 1 foot above and horizontal to a sewer line.

➤ The supply pipe cannot be embedded in concrete and must be wrapped at any point that it passes through concrete.

➤ The minimum size of a supply line is three-fourths of an inch. (If you have three or more bathrooms, figure on a 1-inch line.)

Actually, no pipes are to be embedded in concrete. If your water pressure exceeds 80 pounds per square inch (psi), you'll probably have to install a pressure regulator. Once the main water supply pipe enters a dwelling, it must have an accessible shutoff valve for every unit of the building. This means that a duplex must have one shutoff per unit, a rule that isn't always followed when old houses were converted from single-family homes to duplexes and triplexes. Inside the house, every fixture and appliance should have its own shutoff as well.

Plumbing Perils

Even if your code allows for a three-quarter-inch main water supply line, install a 1-inch line anyway. This will give you plenty of water and allow for future expansion of your plumbing system.

Lending Some Support

Pipes need to be supported as they wind their way through your house framing. Copper pipe needs to be supported a minimum of every 6 feet and plastic every 4 feet. *Every type of pipe has its own support materials!* Plastic supporters go with plastic pipe, and copper supporters go with copper. Don't mix them. These are only minimum support requirements; it's always a good idea to add more support to your plumbing pipes. Pipe support hardware includes straps, hangers, and J-hooks.

In addition to these supports, plastic bushings hold pipes firm as they pass through holes in framing lumber.

Picky, Picky

Most codes call for self-draining, frost-proof hose bibs or sillcocks with built-in backflow prevention. This is a good idea and is worth the small extra cost. Who wants to wake up some morning and discover that an early freeze caught you by surprise and cracked your standard hose bib?

Shower valves must be the antiscald/pressure-balance type to prevent accidental scalding, especially with young children and the elderly, if water pressure drops in the shower because it's being used somewhere else in the house.

Plumbing the Depths

The code addresses only the minimum requirements to provide you with a healthy and safe water distribution and wastewater removal system. You always can go beyond the code by installing extra pipe supports, shutoff valves for distribution pipes, or larger pipes than are called for in any given situation. If the minimum requirements will get you by in reasonable comfort, think of how improved your system will be when you build beyond the minimum.

DWV System: VIP Pipes

It's important to our health that the water coming into our homes is potable and safe. It's equally important that we get rid of our wastewater without it leaking or contaminating our water supply. When repiping your home, you'll have to address your drain pipes and venting requirements. A DWV system will include these elements:

➤ Fixture drains

➤ Fixture branches

Plumbing Perils

Be sure that any vent tee, which connects a fixture's vent to either the main stack or a separate, vertical vent line, is at least 6 inches above the highest fixture served by that vent. This helps to equalize the air pressure in the system, thus preventing siphoning in the traps.

➤ A main soil pipe or stack

➤ Vents

Vent pipes and drain pipes differ in one major respect: The fittings for vent pipes are constructed with sharp angles and virtually no "sweep" or curve to them. Because only air is passing through the vent, the pipes don't need to be concerned about obstructing the flow of wastewater. Drainpipes, on the other hand, must maintain as long a sweep as possible whenever a vertical pipe connects to a horizontal one. The last thing you want is for wastewater to be caught because this angle is too sharp.

Every house is required to have a main vent. This is usually the stack vent, on top of the main stack, that protrudes through your roof. Additional vents that also pass through the roof can be added rather than extending a vent line from an individual fixture to the main stack. You could vent every fixture independently through the roof, but that would lead to a lot of unnecessary holes in your roof.

Your local plumbing code will call for the following requirements in regard to your DWV system:

➤ Trap size

➤ Drain pipe size

➤ Vent pipe size

➤ The maximum distance between the vent and the trap

➤ Construction of wet vents and loop vents

Constructing an accurate DWV system is more complicated than installing water supply pipes. One rule you can count on is to vent downstream from a trap (it doesn't do much good if it's upstream). Trap size is determined by the fixture that the trap serves. We've already noted that plumbing codes have established a system that assigns a value, measured in fixture units, to every plumbing fixture and appliance in your home. The number of fixture units dictates the minimum trap size.

For example, a shower is equal to two fixture units and requires a 2-inch trap. A lavatory, on the other hand, is equal to one fixture unit and requires a 1¼-inch trap. These are only minimums, so there's nothing stopping you from installing the next-larger size of trap and drainpipe.

Drainpipe sizes are also determined by the number of fixture units served. The larger the pipe, the more units it can handle. Toilets always use a 3-inch drainpipe or larger.

Gas Pipe

I don't recommend that you mess around with gas pipe at all (again, your code might not allow you to). Gas pipe comes with its own rules, including these:

➤ You must pass a pressure test.

➤ You may install only iron or steel pipe, unless otherwise approved (such as certain stainless steel tubing products).

➤ You must install a shutoff valve for each appliance.

Gas pipe installation is a good job for a licensed plumber.

Plumbing the Depths

The greater the distance between a fixture and a hot water tank, the longer it takes to get hot water out of the tap if no hot water has been drawn into the fixture for a while. Between draws, the water in the pipe can cool, and you have to empty all that water first before you'll get anything hot. Some large homes get around this problem by installing a circulator pump that provides a constant supply of hot water. I didn't discover this until I asked a plumber why the huge, old Tudor house I grew up in always had hot water available immediately at every fixture, but my smaller bungalow did not.

The Least You Need to Know

➤ A major repipe, or replacement of most of your plumbing system, is a once-in-a-lifetime job for most homeowners (unless you keep moving).

➤ A good pipe layout will save you time and help you avoid installation problems.

➤ If doing a complete repipe all at once isn't possible, you can do it in stages and still keep the water running through the remaining old pipes.

➤ Designing and installing a DWV system is as much or more of a task than installing the water supply pipes.

INDOORS?!

Chapter 20

Adding a Bathroom

In This Chapter

➤ Plan first, build second

➤ Plumbing is just one phase

➤ Know your measurements

➤ Pick an easy location

➤ The extras add up

Bathrooms pop up everywhere in a house. I've seen them built off dining rooms and entryways, squeezed inside former closet space, or on third floors as well as basements. Some of these locations leave something to be desired, such as right off a formal dining room. Simply because you have a ready space available doesn't mean that you shouldn't keep looking for somewhere more appropriate.

Some locations will be more of a challenge than others. A sloped third-floor ceiling might have to be broken out, and a dormer might have to be added. Basement bathrooms can have drainage issues if there isn't enough slope for the drain and waste pipes. The easiest bathroom to add is one that's close to an existing one. Back-to-back bathrooms aren't uncommon and are a good use of a plumbing system.

Building a new bathroom brings in other work as well. Walls must be framed and finished, floor coverings must be installed, and wiring and electrical fixtures must be installed. Natural light isn't necessary, but it's always welcome through a window or skylight, and ventilation from an exhaust fan is a must. Your first step is to narrow down your choice of locations for a new bathroom and then decide which ones are the most logical. With any luck, you'll also pick the easiest one to install your plumbing.

Our National Motto: More Bathrooms

We are a nation that loves our bathrooms. A couple generations back, one-bathroom houses were the standard. Slowly, master bathrooms crept in and then a half-bath or a powder room. Now a finished basement wouldn't be a finished basement without its own bathroom. Having enough bathrooms to go around so that no one has to wait is neither wasteful nor indulgent. Most bathrooms take up relatively little physical space compared to other rooms in a house. If you spend $10,000 on a new bathroom (and you can spend a lot less by doing the work yourself and buying modest but decent fixtures) and you or future homeowners get 20 years use out of it (you can certainly get more), it cost less than $1.40 a day. Who can possibly call that indulgent?

Pipe Dreams

A two-bathroom house is pretty much the minimum expectation in new housing. A second bathroom added to a one-bathroom house will almost certainly pay back its cost when the house is sold. The additional bathroom also will offer you a lot of convenience in the meantime.

Various social observers claim, rightfully, perhaps, that we communicate less these days and that family life isn't quite what we believed it to be in the good old days. Negotiating who gets the bathroom and when they get it isn't the best means for becoming a more communicative culture, though. Add the bathroom first, and then talk to your kids about the meaning of life.

First, you'll need an overview of the project and some measurements.

The Big Picture

Bathrooms are more than just an assembly of pipes and fixtures. This is a project that has to be planned. You don't want cabinet doors bumping into toilets before they fully open, and you don't want a light fixture that's poorly placed. When you've decided on a location for your new bathroom, it's time to pull out the paper and pen or computer design program, and start sketching and figuring. You'll want to know answers to the following questions:

➤ Will the space accommodate the fixtures, including a full bath tub? Will I have to cut back and install a shower only?

➤ How can the fixtures be situated for the easiest plumbing hookups?

➤ What's the best way to light the space?

➤ Is the proposed setup convenient for the intended users (such as children or elderly adults)?

➤ Will there be enough storage?

➤ What will all this cost?

Kohler Camber Lavatory.

Fixtures come in standard sizes and can be configured in any number of ways in your new space, but it's expected that the sink(s) and vanity will be nearest the door and that the tub and toilet will be farther away. Nontraditional design might be fun, but the next home buyer might not think so.

Consider the Plumbing

You have two major access issues when building a new bathroom:

1. You have to bring hot and cold water to the fixtures.

2. You have to vent the fixtures and install drainpipes so that they'll eventually run into a main vent pipe and the soil pipe or main stack.

The closer you are to an existing bathroom or to the soil pipe, the easier your job will be. Most bathrooms get built with these considerations in mind. Every bathroom configuration is different, so it's difficult to outline a single construction route to follow. You want to connect your waste pipes and drainpipes in the most efficient manner possible to the main stack. *Efficient* also means

Pipe Dreams

If your new bathroom space is very limited, consider using a bi-fold door instead of a standard swing-in door with butt hinges. This will resolve any clearance problems with fixtures or counters inside the bathroom that would be caused by a door swinging inward.

with the least disruption to your existing walls and floors. It's one thing to tear them up in your new bathroom, but it's quite another to do so in finished rooms.

Fixtures and Space Demands

Not only do fixtures take up a certain amount of physical space, but you also must maintain certain minimum distances between fixtures for safety and efficiency. The following table lists these sizes and distances:

Bathroom Measurements You Need to Know	
Bathtub	18-inch minimum distance to other fixtures
Shower stall base	32-by-32-inch minimum area
Toilet	15-inch minimum distance from the toilet flange to the side walls; 1 inch between the toilet tank and the back wall; 30-by-30-inch minimum free floor space in front of the toilet
Door/entrance	32-inch minimum distance
Sink	12-inch minimum distance from the center of the sink to any side wall; 21-inch clear floor space in front of the sink

Your space might allow for only a corner sink, a shower stall, and minimal storage. On the other hand, you might have a huge space available and can go all out with double sinks, a separate shower and soaking tub, a bidet, and a storage closet. You have to maintain only the minimum dimensions shown in the table, but can always be more generous with your spacing.

Plumbing Perils

If your plan is to upgrade an existing bathroom and add a second one, put off the upgrading until the new bathroom is completed. You always want one working bathroom in operation while you're working on another one.

What's All This Going to Cost?

Shop around at both plumbing suppliers and home improvement stores when pricing fixtures and building materials. Be sure that you're comparing the same products or ones that are equivalent. Inquire about discounts from plumbing suppliers if you buy all your fixtures and pipe from them. If you're hiring some of the work out, get a breakdown of labor and material costs from your contractor.

Code Issues

A new bathroom will introduce you to a couple other building inspectors besides the one who looks at your plumbing. You will need an electrical inspection and an inspection for the framing and flooring, too. Code requirements include these:

➤ A GFCI receptacle

➤ One switch-controlled light

➤ An exhaust fan or operable window

➤ Two-by-two framing, with each stud secured with two fasteners at each end (two-by-six framing is also acceptable)

➤ In basement installations, a *bottom plate* that is *treated lumber*

➤ Insulation

➤ Pressure-balancing mixing valves for the shower (depending on code)

Find out from your building department what kind of plans are required for a permit. The more detailed the plans, the easier it will be for both you and the building department to discover any flaws before you start building.

Windows and Ventilation

I've already written about the importance of a good exhaust fan and won't belabor the point again. An operable window provides ventilation, but it doesn't suck out steam and odors. Privacy is another concern and could determine the size of window you install. The style of the window should match your existing windows, especially if you have an older, historic house. If your neighborhood has building and design covenants, you'll probably have to match your other windows.

An existing window might have to be moved, or the size of the opening might have to be changed. This will require you to repair and fill in your exterior siding, a chore you'll have to do off extension ladders if your bathroom is on the second or third floor. This can be an extensive repair if you have regular wood, vinyl, or aluminum siding. You don't want to simply install a series of small pieces to fill in if you're swapping an existing large window for a smaller one. Well, you might want to patch up the siding this way, but it will look tacky and draw attention to the window. You should duplicate the existing siding pattern of random lengths.

What's That Thingamajig?

A **bottom plate** is the bottom horizontal framing member of a stud wall. **Treated lumber** has a chemical preservative to protect it against rot. It's used against basement floors because of potential exposure to moisture.

Plumbing Perils

Some old homes have a window right above a bathtub without a shower. As showers got added in later years, these windows ended up being a liability and were easily subject to water damage. Showers and windows, especially wood windows, don't mix, so leave them out of your new bathroom shower.

Patching in shingles, stucco, and, to a lesser extent, bricks, is often less extensive than patching in wood siding.

Framing

Skip this section if you already have an existing room for your new bathroom. Otherwise, follow these simple rules when you do your framing:

Pipe Dreams

An easy way to visualize your fixture locations is to outline them with masking tape. Blue tape is more visible than the beige-colored masking tape.

➤ Purchase straight lumber without any bows or cupping.

➤ Use a carpenter's level when you're building your walls—don't depend on "eyeballing."

➤ Measure twice, cut once.

➤ When building over a concrete basement floor, use construction adhesive on the underside of the wall's bottom plate in addition to fasteners.

The minimum construction standards call for two-by-four wall studs, but two-by-six "wet walls" (the walls that contain the pipes and drains) ensure plenty of room to do your plumbing tasks. Install plenty of cross bracing between the studs to secure the shower pipe and waste pipes, where needed. If you have to remove any existing plywood or oriented strandboard (OSB) subflooring or underlayment, remove a generous enough section so that you can do your work without being hampered. When removing a section of subflooring, follow these steps:

Plumbing Perils

Be careful when working around exposed floor joists! Everyone has a story about a worker putting a foot through the ceiling below when accidentally stepping between the joists. Always lay scrap pieces of plywood across the exposed joists so that you'll have something to walk on.

➤ Mark it off with a T-square and a pencil or crayon.

➤ Use an electric circular saw with the blade set to the thickness of the plywood.

➤ Use the nails that secure the plywood as a guide, and cut so that the remaining section of plywood is still secured to a joist.

When you reinstall the piece of plywood that has been removed, simply nail and glue a length of two-by-four or two-by-six onto the exposed joist so that you'll have a nailing surface.

This Thing's Heavy!

Cast-iron tubs and jetted tubs are not only heavy on their own, but they're really heavy when they're filled with water and, well, you. You should figure on doubling up the floor joist. These "sister" joists should be secured to both rim joists, but sometimes that isn't possible. Discuss the problem with a contractor or your building department for a satisfactory solution.

Insulation Needs

Fiberglass insulation with a vapor barrier is the usual choice for insulating exterior walls. You might consider insulating all your bathroom walls for the sound-dampening value—falling water, both our own and that from the pipes, is noisy and is easily heard through walls lacking insulation. It's a good idea to staple a clear plastic vapor barrier across all the framing after the rough-in work is completed to protect it from any moisture damage.

Electrical Requirements

The electrical layout for a bathroom is fairly simple. One cable will control the GFCI receptacle, and one will control the lights and the fan. A third cable will supply power to any heat source, such as a baseboard heater. Under some conditions, the GFCI can be run off an existing GFCI, and the lights and fan can tap into an existing circuit if it can handle the extra loads. Follow this simple rule: Don't mess with your electrical system unless you know what you're doing! That's why there are electricians in the world.

You might have to run one or both of your cables back to your service panel and install new circuits for the bathroom. You'll be drilling five-eighths-inch holes (minimum) in the studs and joist to run nonmetallic sheathed cable that contains the hot, neutral, and grounding conductor required by the National Electrical Code. The cable will terminate at electrical boxes (either metal or plastic) and connect with switches, the GFCI receptacle, the fan, and lights. The boxes are nailed to the framing and stick out far enough from the edge of each stud or joist that they will line up flat with the finished drywall.

Anywhere the electrical cable passes through the framing, you should nail on a protective steel nailing plate, the same as you do with pipes. Give yourself a good 8 inches or so of extra cable coming out of the box so that you'll have plenty to work with when you do your final connections.

> **Pipe Dreams**
>
> Electrical boxes come in different sizes. Instead of installing the minimum required, install a bigger box. These are much easier to work with when you have to connect wires to switches, receptacles, and lights.

The electrical rough-in work—that is, the installation of the wires into the boxes—must be inspected and approved before you install the drywall.

It's Cold in Here!

Bathrooms need heat, and an electric baseboard or resistance heater with a fan blower will do the trick every time. If your new bathroom is near a furnace duct, you can cut into it, install an adapter, and run a flexible duct to the bathroom. A vent can be cut into a duct that runs overhead through the bathroom ceiling.

Cutting into a duct will bring some heat, but probably not enough for you to be completely comfortable. Figure on installing supplemental heat when you plan your wiring.

The Difference Between Drywall and Backerboard

This is simple: Use drywall anywhere you're going to paint, paper, or install laminate, and use backerboard for tile surfaces. Greenboard, with its water-resistant paper covering, should be installed in bathrooms. Repeat this mantra: Water-resistant is not waterproof! Greenboard is mistakenly used under tile all the time because it's less expensive than cement backerboard. It never lasts, though, and at some point down the road it has to be removed and the cement backerboard installed. Any surface that's going to get wet and be tiled, including the floor, tub area, and shower stall, should first be covered with backerboard. Sure, it's more of a nuisance to install and costs more than greenboard, but what's the price of pulling it all out and starting again?

Plumbing the Depths

I knew a spa owner who had built a dozen tiled rooms complete with showers, hot tubs, and saunas. Every one of them was built using greenboard, and every one of them had to be torn apart and built again at a huge cost. He was acting as his own contractor and discovered that the greenboard wouldn't hold up. He recouped his costs, but it was a very expensive lesson.

If you haven't hung drywall before, it's not as easy as it looks, especially in small rooms with multiple angles. Drywall hangers charge by the square foot on large commercial jobs, but small jobs will be billed by time and materials. Joints have to be tight, and the boards have to be cut precisely around electrical boxes and pipes. The material is relatively inexpensive, so you can always practice on a sheet or two, but it's well worth hiring out. Poorly installed material is also more difficult to tape and finish properly.

Choosing Cabinetry

Every plumbing showroom and home center carries finished vanities, with and without sinks, that offer simple solutions to your storage needs in a new bathroom. You can often order the cabinet in different finishes and colors and can choose from a selection of tops. Then it's just a matter of carting it home, shoving it into place, and hooking up the pipes. This is much simpler than building a vanity in place, but you might not get the exact dimensions you want with a finished unit.

Unless you're a cabinetmaker, the chances are slim that you'll build your own vanity or storage cabinet. Local suppliers of bathroom vanities can give you the name of a custom builder, or even their own builder if he or she takes special orders. Ideally, you can order a unit to fit your space rather than one that is built and assembled partially on the job site.

Your vanity should be installed level and plumb. Slide it in first to determine the pipe locations. Mark the back of the vanity, pull it out far enough to drill some pilot holes for the water supply pipes and the drainpipe, and then drill from the inside of the cabinet.

Pipe Dreams

The easiest vanity top to install and clean is a cultured marble counter. These are one-piece designs that include a sink. All you add is a faucet. These models are long-lasting and can take a lot of abuse (not that you should be abusing your fixtures).

Plumbing Perils

Most medicine cabinets are set into the wall between two of the studs. When you're framing, be sure to allow for this, and leave a space centered over the vanity. Otherwise, your medicine cabinet will be off-center.

Tiling Thrills

Tiling requires a special saw with a water-cooled blade that's available from most tool rental companies. The tile has to be fitted, cut, spaced, and grouted. Tiles also have to be notched around the faucets and shower arm. The tile is pushed into a bed of thinset, which is a type of mortar. Tiled shower pans require a bed of mortar, a waterproof membrane, more mortar sloped toward the drain, and finally the tile.

Do you really want to do this? I didn't think so. Hire a tile setter. You'll be much happier.

Treat Your Feet to a Fine Floor

We covered floor coverings in Chapter 18, "The Big Picture." Your subflooring should be glued and fastened to the joist with either nails or screws. A glued floor has less chance of squeaking. A basement concrete floor can be dealt with in one of two ways:

1. You can install your finished flooring directly over the concrete.

2. You can raise the floor and lay down plywood.

A raised floor means that you secure treated lumber to the concrete floor, fasten plywood over it, and install the finished floor material over the plywood. This produces a softer and warmer feel to the floor because it's not in direct contact with the concrete, but it's not the greatest idea in the world. There's the issue of vermin or water getting between the boards and making a real mess of things. Stick with a vinyl floor laid directly over the concrete. It's simpler and cleaner. A few washable throw rugs will keep the floor comfortable.

To lay vinyl flooring, follow these steps:

➤ Fill any cracks or low spots in the concrete with a concrete filler; if you're going over plywood, use a floor-patching material such as Fixall until the floor is fairly level.

Pipe Dreams

The advantage of individual floor tiles is that single tiles can be replaced in the event of damage. This benefit is more pertinent in a kitchen or commercial area, but it can apply to bathrooms as well.

➤ If you're installing sheet vinyl, you can either measure the room and cut the vinyl, or lay out a piece of scribing felt (your floor dealer sells this) and make a template.

➤ Cut the vinyl in an open space where you have room to roll it out flat.

➤ Lay the vinyl out in the bathroom to test the fit.

➤ Follow the manufacturer's instructions and those of the adhesive manufacturer for installing the vinyl.

Vinyl tiles are installed one at a time, of course, and allow for some creative floor patterns.

Fixtures and Faucets

These can be real budget-busters, so consider what you really need and can live with day to day. There's no point in choosing a line that will always feel like a compromise. You can't go wrong with white fixtures, but if you choose a special color, be sure that you *really* like it because you'll be looking at it for a while, too. After all, there are no returns or exchanges after your fixtures are installed. Pick brand names for their warranties and parts availability.

If you have lots of storage room and don't need a vanity with its own storage, consider installing a pedestal sink. It can lend a certain artistic and less utilitarian look to your bathroom.

Bathroom Odds and Ends

When the walls are up and painted and the fixtures, wiring, and flooring installed, you'll still have a lot of finishing details, including these:

➤ Wood trim and casings

➤ Towel racks

➤ Clothes hooks

➤ Mirrors

➤ Globes and bulbs for the light fixtures

The single best tool for cutting wood trim is an electric miter saw. It will cut faster and at more precise angles than you could ever do by hand. All nails should be set and the nail holes filled prior to painting. If you're staining or clear-coating the wood, fill the nails with a putty stick, available in colors to match your woodwork, *after* the finishing.

Plumbing Perils

Wood putty is also used to fill nail holes, but it doesn't stain the same color as the surrounding wood. If the wood has been sealed first and then the putty used and sanded smooth, the entire area around the putty will look splotched when the stain is applied. If you use wood putty, fill the nail holes, sand the wood, and then apply any sealer.

When ordering a wall mirror, your measurements must be accurate if you're installing it yourself. Mirrors are cut by glaziers the same way any pane of glass is cut. Try to attach one that's a half-inch too big onto your wall, and you're going to have a problem. Glaziers install mirrors as well as glass, and you might want to hire this out if the mirror is a large one. The experts will measure and cut at the job site for a perfect fit.

Simple Installation Rules

You don't want to go to all the trouble of installing a new bathroom and then end up having problems because of some foolish mistakes. Some bathroom construction rules are universal, including these:

➤ Be sure that all your fixtures are level when you're installing them.

➤ Install water hammer arrestors. You won't know until it's too late whether you need them, so go ahead and put them in.

➤ Cut the branch lines on the long side so that you have plenty of pipe to work with after the drywall has been installed.

➤ Insulate any exterior walls to avoid freezing pipes during the winter.

Leveling isn't just an aesthetic consideration. It also helps ensure that fixtures are properly sealed, especially toilets.

Basement Bathrooms

The higher the bathroom is from your sewer line, the more it has gravity on its side when it needs to drain wastewater. Basement bathrooms aren't so lucky and have to be carefully planned. Instead of running your drainpipes and waste pipes through your home's framing, you'll be chopping up your concrete floor and running a vent pipe up to the attic. It might be necessary to install a toilet with a sewage injector if the closet bend would otherwise fall below the sewer pipe.

Basement installations can be complicated. It's a good idea to consult or hire a plumber for this job.

A basement installation.

The Least You Need to Know

➤ It's hard to go wrong adding a second or even third quality-built bathroom to your home.

➤ For ease of construction, locate your new bathroom as close to existing plumbing as you can.

➤ Do plenty of shopping around to look at different fixtures and faucets; you'll get a better idea about prices at the same time.

➤ If you don't want to do the entire bathroom addition yourself, there are plenty of tasks you can do while only hiring out the specialty or more complicated work.

Chapter 21

WHERE'S THE KITCHEN?

Kitchen Remodeling

In This Chapter

➤ Consider your lifestyle first

➤ Working around your existing kitchen

➤ Coming up with a design

➤ Plumbing choices

A full kitchen remodel, or the construction of a new kitchen in a house under construction, is a big job. The plumbing is pretty straight forward, and because the kitchen is closer to the main water supply pipe than a second-story bathroom, most of your pipe runs will be horizontal through the basement or crawlspace with short vertical runs into the kitchen. This is easier than going way up through the walls, especially if you're remodeling an existing house and have to cut into plaster or drywall.

The least expensive approach, as far as the plumbing goes, is to use your existing pipe and the current locations of the sink and dishwasher. Simply upgrading the sink and the appliance is relatively simple. Gutting out the existing galvanized pipe, relocating a sink, or adding a second sink or dishwasher is more of a job.

A total remodel involves a lot of other decisions besides plumbing fixtures and appliances. A traffic pattern must be established, counters must be selected, cabinets must be installed properly (you don't want doors and drawers colliding), and lighting must be installed, among other tasks. This chapter goes over the general kitchen remodeling process. A kitchen is typically the most expensive room of the house, but with a little knowledge and planning, you won't have to rob your retirement funds to pay for it.

An Overview

The kitchen is the modern equivalent of the cave or hut, the first shelters that distinguished us from the other mammals running around. Instead of gathering around the fire watching the catch of the day cooking over an open flame, we now sit around a center island with a built-in grill top and an online Internet appliance in the corner searching for a recipe for salmon filet. Kitchens have gotten bigger than ever and are the true gathering place in our homes.

At its most basic, a kitchen is a room for storing and preserving food, cooking and preparing meals, and tucking away cooking implements and our table service. We could get by with a cheap stove, an ancient refrigerator, a garden hose and a shower basin for washing dishes, and cardboard boxes for storage, but we're not willing to live like that when we don't have to. A well-designed, comfortable kitchen is one of the treats of American housing, and we should take every advantage of it.

Kitchen design is based on the answers to all kinds of questions, including these:

➤ What's your cooking style?

➤ How tall are you?

➤ How many meals will you prepare each day?

➤ How many people will be preparing food and cooking at the same time? How many will be cleaning up at the same time?

➤ Do you prefer a lot of natural light?

➤ Do you want an open kitchen or a more private space?

➤ What will you do in your kitchen besides cook and eat?

Pipe Dreams

Home shows are great ways to get a look at the latest and greatest in consumer products geared toward remodeling. Plan on spending several hours snooping around. Grab any brochure or business card that may be of some use to you later. Take a notebook with you as well to record other information pertinent to your projects.

If you live alone and mainly use a kitchen to microwave leftover Chinese take-out and refrigerate your microbrews, you don't need much of a kitchen at all. On the other hand, if you and your partner are gourmet wannabes who buy imported French salt and organic red peppers, you're going to want it all: two or three sinks, a marble counter for pastry making, and a six-burner commercial gas range. Your physical stature and that of any other regular cooks will affect your kitchen design. For instance, you can install some counters lower than the standard 36-inch height.

It's Getting Crowded in Here

Everyone ends up in the kitchen, it seems, especially during parties. A kitchen can be welcoming while still being efficient for its cooks. Most kitchen planners design around a work triangle or the shortest walking distance from the refrigerator to the stove or range, then on to the sink and back to the refrigerator. Measuring from the center of each area, the maximum total distance should be no more than the magic figure of 26 feet. No single side of the triangle should be longer than 9 feet or shorter than 4 feet. Designers consider this triangle and its measurements to be a sacred symbol of their profession and have declared that no major foot traffic should ever cross any of its sides or legs.

Plumbing the Depths

According to the National Kitchen and Bath Association, the work triangle resulted from research done by the Small Homes Council of the University of Illinois and the Home Economists from the Agricultural Research Service in 1949. It was meant to simplify laboring in the kitchen by connecting the three major work areas in a serviceable design.

This is a fine ideal, but some kitchens just won't meet all the criteria of the work triangle, although it's good to work toward it in your design. Work aisles—the space between counters—should be at least 36 inches when possible. Other helpful measurements that will result in a workable kitchen space include these:

➤ Cabinets installed over counters should be at least 12 inches deep and 30 inches high.

➤ Cabinets under a counter should be at least 21 inches deep.

➤ Allow 15 to 18 inches of clearance between the counter and the bottom of wall cabinets.

➤ A dishwasher should be installed within 36 inches of a sink.

Remember, these are measurements you should try and work toward, but they're not absolutely sacred!

Electrical Req-wire-ments

Your local electrical code lays out the rules for kitchen wiring. Because it's such a heavy user of electricity, there are a number of regulations to follow. Most local codes are based on the National Electrical Code (NEC), a virtual tome of electrical minutia. The NEC states the following regarding kitchen wiring:

➤ Each major built-in appliance must have its own dedicated circuit. (The refrigerator is an exception.)

➤ At least two 20-amp small appliance circuits are to be installed, and each is to have ground-fault circuit-interrupter (GFCI) protection.

➤ These same small appliance circuits cannot be used for built-in lighting, exhaust fans, or built-in appliances.

➤ The circuit serving the disposer does not have to be GFCI-protected.

➤ The refrigerator receptacle may be supplied by a separate 15- or 20-amp circuit, but this is not required.

Built-in lighting will be on its own circuit, of course, and this may be shared by lights in another room. The key issue in a kitchen is the number of dedicated circuits to individual appliances and to small, plug-in appliances. Although no absolute ventilation level is mandated, install a big fan. You want the odors and grease to take a hike outside, not hang around and set up camp on your kitchen walls, cabinets, and ceiling. Exhaust fans come in three basic flavors:

Plumbing Perils

Keep a small fire extinguisher in your kitchen where it will be readily available in case of a grease fire. Don't hang it too close to the stove, however.

➤ Fans that recirculate and are not vented

➤ Vented fans with internal motors

➤ Vented fans with roof-mounted motors

Fans that are not vented simply pull in steam and odors, and pass them through a charcoal filter before recirculating them back into the room. They're better than no fan at all, I suppose, but not by much. Vented fans force odors outside where they belong. A unit with a roof-mounted motor is much quieter inside your kitchen.

Lighting

In a kitchen, you need general ambient lighting and task lighting so that you can see what you're chopping. Sometimes overhead lighting is sufficient for your cooking chores, and sometimes you need smaller, specific task lighting such as a small fluorescent fixture installed on the bottom of a wall cabinet. Better yet, at least during the day, is an abundance of natural lighting through windows or skylights.

Looking Ahead

It's great to live in the moment, and one can argue that this is all we really have anyway. Still, tomorrow seems to repeatedly rear its head, so we ought to keep it in mind when we do any remodeling. You may be the unfettered couple today, but if you expect to have a family, you might reconsider whether you can convert half your finished basement to a home brewery instead of a future playroom. If you want both, is your current house the right one for you? Will you stay in this house long enough to recover your remodeling costs? Given how often Americans move, these are not idle questions.

A good kitchen remodel will often return most of its cost when you sell your house, but there are no guarantees. Your house may be a home first, but it's a good idea to consider your remodeling dollars as an investment as well. There's no way to foresee what materials, finishes, and trends will be popular in the future when you decide to sell, but good taste never goes out of style. That said, avocado-tone appliances were once *very* popular, and now you can't give them away.

Pipe Dreams

Put a realistic time frame on how long you expect to stay in your home. If it's a temporary residence that you'll be moving from in two or three years, it makes little sense to build a top-of-the-line kitchen that you'll barely enjoy. Bring it up to code, make it presentable, and let the next owner do the big remodel.

A Computer in Every Room

Up-and-coming computer appliances will be the size of a desk phone and won't need any special space allotted to them. For that matter, they'll be a combination phone and Internet device that will hang on the wall or take up a small area on a counter. You might want to allow a little more space for a laptop near your phone jack just to hedge your bets. The idea of grabbing all our programs off the Internet as we need them sounds like a huge loss of personal control to me.

Floors

This is simple: Don't install carpet, think twice about wood, you can always redo vinyl, and ceramic tile is essentially permanent. Wood floors in kitchens are very popular right now, but they show every scratch regardless of the finish used to protect them. They stay scratch-free only if you're a single, childless, petless adult who walks around in stocking feet and has no kitchen chairs.

Refinishing wood floors is a big deal, and it's nothing you want to have to do in a kitchen. If you want the look of wood, consider laminate flooring such as Pergo brand floors. Vinyl is very versatile, available in tile and sheet form, and a good value

Plumbing Perils

Every floor covering is subject to damage. Drop a heavy enough iron pot on a tile floor, and you're going to crack something. Laminated floors are pretty scratch-resistant, but don't view that as a license to abuse them. Treat and maintain your floors according to the manufacturer's instructions, and your floors will serve you well.

for the money. Ceramic tile will last forever, but it's both cold and hard and always at war with bare feet in cold climates. Each material has its own installation requirements and maintenance issues, so look at them all carefully before making your decision.

Lots of Storage

Cabinets are boxes with doors on the outside and shelves on the inside. Unfortunately, they can be surprisingly expensive boxes depending on your taste and willingness to spend. They can account for almost half the cost of a new or remodeled kitchen. Cabinet styles include these:

➤ Wood with stained and clear-coat finishes

➤ High-density particle board with plastic laminate or painted finish

➤ Doors with raised panels or glass panels

➤ Brass or color-coordinated hardware

Pipe Dreams

Instead of allowing the soffit area above your cabinets to be empty space, design your kitchen to run the cabinets all the way to the ceiling. This can be great storage space for extra canned goods, rolls of paper towels, or cookware that's used only occasionally.

Storage includes cabinets and drawers as well as open shelving, wine racks, and overhead racks for hanging pots, pans, and large cooking utensils.

Even the smallest kitchens have unused space available for storage. A cutting board can always hang on the outside of a cabinet, or one-piece wire shelving normally meant for a clothes closet can be installed on the inside of a pantry door to hold herbs and spices. Take a walk through a store that specializes in storage containers and shelving, and get some ideas.

Reuse, Refinish, Resurface

The original cabinets in many old homes were sometimes built on site and would be prohibitively expensive today. Rather than ripping them out, consider whether you can reuse them. Original cabinets can be:

➤ Cleaned up and repainted

➤ Stripped and refinished

➤ Resurfaced

Repainting is the easiest option after all the painted surfaces are sanded and deglossed (watch out for lead paint). Old varnished cabinets can look like new after being stripped, sanded, and refinished. Resurfacing is the application of another material, such as plastic laminate, over the existing wood. Typically this choice is coupled with new hardware. For a change of style, you can keep the cabinet frames, replace the doors and drawers, and install mechanical drawer glides.

Kitchen Designers

In an era of specialization, you can even hire someone specifically certified to design your kitchen. Although an architect or talented builder can do the job, the National Kitchen and Bath Association (www.nkba.com) actually certifies designers who meet their stringent requirements and testing. Their certifications include these:

➤ Certified kitchen designer (CKD)

➤ Certified bathroom designer (CBD)

➤ Certified kitchen and bathroom installer (CKBI)

On the other hand, many home centers offer in-house planning and drafting assistance with their computer drafting software. Just bring in your kitchen's dimensions, and they'll run them through various cabinet configurations (based on the lines of cabinets they sell). These won't be elaborate plans, but they're a start and will give you an idea of material costs.

Pipe Dreams

Your old fixtures, appliances, and cabinets have virtually no resale value, but they can be reused if carefully removed. Used kitchen cabinets often end up reinstalled in basements and garages for storage. Working appliances can be donated to some nonprofit outfits (find out who will do a pickup). In farm or ranch country, old bathtubs are sometimes used as watering troughs for livestock.

Putting In the Plumbing

Your plumbing layout for your kitchen will depend on the number of fixture units being installed. At a bare minimum, you'll have a kitchen sink. At a maximum, you'll have two or possibly three sinks, an on-demand hot water dispenser, and a dishwasher. In addition, you might install a refrigerator with an ice dispenser that will require tapping into an existing cold water supply pipe.

A dishwasher is rated at two fixture units. A kitchen sink is one fixture unit. It doesn't make any difference how big the sink is because it has only one pair of hot and cold water supply pipes connected to it. An ice dispenser doesn't count as a fixture unit, nor does a hot water dispenser under the kitchen sink.

The sink, dishwasher, and food disposer are integrally connected and can almost be thought of as one grand combo unit. They all share a single trap and drain, and the disposer is almost like a specialized sink drain itself. About the only times you see a standalone sink anymore without these two appliances is in older homes and apartments.

Plumbing Perils

Be sure that any pipes running inside an exterior wall are well insulated. You don't want these to freeze during a cold snap and possibly burst. Wrap them in pipe insulation, and pack the wall cavity with fiberglass insulation as an additional precaution. New construction will require fiberglass insulation anyway.

It's recommended that you run three-quarter-inch copper pipe (type M) as your main distribution pipe and half-inch pipe for the branch pipes that supply individual fixtures. Use three-quarter-inch pipe for any branch line supplying more than one fixture. This will provide plenty of water for your dishwasher and sink(s). The supply tubes running off the branch pipes should also be at least half-inch size.

Thrown for a Loop

An island sink doesn't have a nearby wall to install a vent pipe. A loop vent will be required. Local building codes are very specific about loop vent installations, so be sure that you or your plumber's plans are approved by your local building department.

Island sinks are typically smaller than the main sink. Some homeowners install deep bar sinks and use them mainly for rinsing food and hand washing. It's unusual to need more than one large sink for dishwashing.

Speaking of Venting

All your kitchen drain lines must be vented, even if your current kitchen isn't vented properly. You'll have to bring it up to code. This will include removing old S-traps that are no longer legal. You'll be installing P-traps with any new fixtures.

Use What You Have

Whenever possible, tie in your new work to the existing waste and vent pipes. If you have old galvanized pipes, remove them and install new copper or plastic. Run your

What's That Thingamajig?

A **riser** is a vertical section of pipe that connects a fixture to the water supply.

pipe as efficiently and straight as you can, with the fewest number of bends. An existing steel or iron drain line can usually have a new plastic drain pipe tied into it by cutting a section of the old pipe out with a reciprocating saw and using the proper couplings.

Run your branch lines from the basement or crawlspace below and up and into the wall behind the main sink. These vertical risers should extend 12 inches or so up from the wall plate. The *risers* will be fitted with male-threaded adapters. Shutoff valves then will be screwed onto the adapters.

Picking a Faucet

These days, you'll almost always pick a faucet that uses cartridges or ceramic disks rather than compression faucets with rubber washers. Look at riser faucets that are tall enough to accommodate washing out large pots and pans. Decide whether you want a spout that doubles as a pull-out sprayer or a spout that's separate from the sprayer. You may even want a faucet with a built-in water filter.

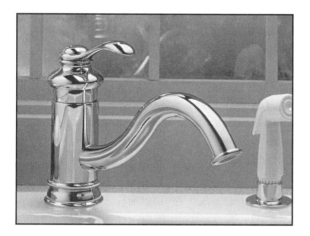

Kohler Fairfax Kitchen Faucet.

Dishwashers

You can have as many options as you want here, from plain vanilla to multiple wash cycles, adjustable racks, and built-in food disposers. The plumbing installation will be the same for any model, with some changes in the actual water supply hookup. The hot water will be supplied by a flexible tube or copper connected to the hot water supply pipe under the sink.

Structural Changes

Many old kitchens had breakfast nooks and walls between the kitchen and the dining room. There's nothing like removing a wall or two to really open up a kitchen. Tearing down a *non-load-bearing wall* is not big deal, but a *load-bearing wall* must be replaced with an alternative means of supporting the joist overhead. Typically, this means installing posts and a beam. If you're not sure what kind of wall you're dealing with, put your saw and sledgehammer away until a contractor or architect can take a look at it.

What's That Thingamajig?

A **non-load-bearing wall** is simply a partition and isn't holding anything up.

A **load-bearing wall** is one that helps support the house. It's critical to the safety and integrity of the framing and usually runs perpendicular to the floor joist.

If your kitchen is attached to a back porch, consider enclosing this and using it as finished kitchen space. You shouldn't have any problems with setbacks (clearances involving the lot lines), and you can pick up quite a bit of space. The porch area will have to be brought up to insulation and structural standards for interior living space.

If you have enough room between your house and the neighbor's, you can knock an exterior wall out a couple of feet or more, and give yourself all kinds of room for a dishwasher, wider counters, and a greenhouse window. Building codes determine how close you can build to your lot lines, and your building department will approve your expansion plans only if they stay within these guidelines.

Plumbing Perils

Neighborhood convenants can restrict exterior alterations to homes within that neighborhood. The alterations that are allowed normally must pass committee approval. Historic neighborhoods often have these types of restrictions.

Refrigerators

Side-by-side refrigerator models are generally more expensive than those with an upper or lower freezer unit, but they also come with ice dispensers built into the freezer door and just look really sleek. The ice and cold water dispenser will require a connection to a cold water supply pipe, typically done with a saddle valve, and a supply tube between the valve and the refrigerator. Soft copper works well here and allows you to move the refrigerator for periodic cleaning behind it. Refrigerators are pretty economical to run, with the national average being around nine cents per kilowatt hour.

Gas Range and Cooktop

Many cooks prefer gas because of the control it offers over heating temperatures. Most gas ranges cost more than their electric counterparts. A gas range will require gas piping, of course, and an electrical hookup to run the electronic ignition and clock. A shutoff valve should be installed near the range, with a flexible connection hooking up the gas supply with the appliance.

Commercially inspired models are available for home installations. They're expensive and come with extra burners and cooking capacity. Even a properly adjusted gas range gives off some exhaust gases when in use, and it's best that they be vented outside rather than having to depend on a ductless system of ventilation.

The Least You Need to Know

➤ Before doing any kitchen remodeling, consider how you and your family use the kitchen; your usage should help determine its design.

➤ Whenever possible, reuse any existing plumbing that's up to code and that fits with your new sink(s) and appliances.

➤ Second sinks can require special venting procedures; consult your local code before doing any installations.

➤ Given their expense, it pays to look at a lot of cabinet and counter options before making a final decision.

➤ A good-quality vinyl floor will give you the most bang for the buck out of many flooring choices.

➤ Hire a plumber to install any pipe for a gas range and cooktop.

Outdoor Plumbing

The principles of outdoor plumbing are the same as those for indoor plumbing, except that you don't have to worry about a DWV system. This is strictly one-way water supply. Drainage might become an issue if you have an unusually wet yard or if water accumulates in certain areas due to soil conditions. This water typically is directed into a so-called French drain, a hole or ditch dug several feet deep and filled with rocks or stones.

Most of our outdoor watering needs are directed at gardening and washing (cars, boats, dogs). You should have at least one hose bib or sillcock at both the front and the rear of your house. If you're feeling particularly energetic, you might install more. A long hose can make up for fewer hose bibs, but special landscaping or yard designs might benefit from more localized water access.

In addition to garden hoses, which everyone has, outdoor sprinkling systems are a common outdoor plumbing project. These systems use plastic pipe—although old systems used metal pipe—and require careful planning to ensure proper distribution of water to gardens and lawns. New sprinkler systems can be broken up into zones, and each can be separately controlled by a timer. On top of that, rain sensors detect whether the system is needed at all and will shut it off when necessary. There are some installation rules and precautions to follow for outdoor plumbing, and we will discuss them in this chapter.

It Starts with a Hose

Even if you have a yard full of gravel, plastic, and potted plants, you'll most likely have an outdoor faucet on two sides of your house. Various terms are used to describe outdoor faucets, including these:

➤ Sillcock

➤ Hose bib

➤ Bib faucet

A hose bib is a shorter version of a faucet, used most often in a house to supply water to a washing machine. It's attached directly to a water supply pipe as opposed to a sillcock, which comes attached to a section of thicker pipe. For the purposes of this book, we'll stick with sillcock, which seems to be the most common plumbing term for an outdoor faucet. (Even the supply catalogs are inconsistent in their use of the terminology.) A homeowner's main concern with a sillcock and any outdoor plumbing is whether it will withstand freezing weather (anyone living in a tropical climate can skip this section).

If you haven't got one now, you might consider a frost-proof or frost-free sillcock.

Pipe Dreams

Outdoor plumbing does mean one thing you don't do much of indoors (except in the basement), and that's digging. Unlike your main water supply pipe, your sprinkler pipe does not have to be below the frost line because it will get drained before cold weather sets in.

Ice in the Pipes

Cold, winter weather means freezing temperatures. Any water that isn't drained from exposed pipes—those that are above the frost line—will change to ice and expand, possibly bursting the pipe. For this reason, homeowners should drain the pipes leading to their outdoor sillcocks when the weather begins to get cold and they no longer need to use an outdoor faucet or hose.

To drain a sillcock, follow these steps:

➤ Find the shutoff valve for each sillcock.

➤ Shut off the valve.

➤ Open the sillcock and drain out any water in the pipe.

➤ When all the water has drained out, close the sillcock.

Plumbing Perils

Be sure to drain any garden hoses and store them in a garage or basement for the winter. Never leave a hose attached to a sillcock during freezing weather because water in the hose at the female end that attaches to the faucet can expand and burst the hose and possibly the faucet.

Unless you drain your pipes religiously, it's a good idea to install a frost-proof sillcock, especially for those sudden drops in temperature that can catch you off-guard. A frost-proof sillcock is self-draining, so there's no danger of freezing water expanding and damaging it. Still, if you have an existing shutoff valve for an individual sillcock, you should retain it even if you're installing a frost-proof model.

Installing a New Sillcock

Frost-proof sillcocks come in various lengths, from as little as 6 inches to as long as 30 inches, to accommodate different installation locations. This length of pipe is thick and made to withstand freezing temperatures. When you turn on the sillcock, this pipe fills with water; when you shut it off, the end of the pipe seals where it connects to your water supply pipe. This way it never has any water in it that can freeze.

The type of pipe that you have and the location of the shutoff valve will determine the installation. Old homes often have threaded galvanized piping, and some codes may not allow you to attach plastic or copper pipe to galvanized (always check first).

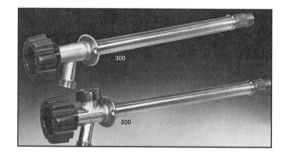

Frost-proof sillcocks.

Mansfield Plumbing Products, Inc.

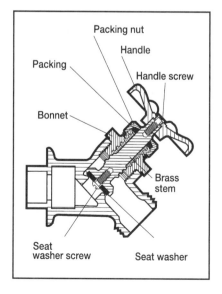

Typical sillcock.

To install a new sillcock, follow this procedure:

➤ Shut off the water at either the individual or the main shutoff valve, and open the sillcock to drain off any water from the pipe.

➤ If you have galvanized pipe, unscrew the sillcock from your house—there are two screws that secure its flange to your siding. Secure the shutoff valve with a pipe wrench, and remove the section of pipe between the valve and the sillcock with a second wrench.

➤ If you have copper or plastic pipe, cut only enough pipe to enable you to remove the old sillcock, but leave enough to attach the new frost-proof model. (This will vary depending on the length of the new sillcock.)

➤ Pull the sillcock and pipe out and away from your house, and push the new sillcock through the old opening.

➤ Install a galvanized coupling first and a transition fitting for galvanized-to-copper pipe second onto any existing galvanized pipe. If you have plastic pipe, install the appropriate threaded transition fittings. (Copper to copper might not need a coupling, depending on the diameters of the pipes.)

➤ Clean all pipe ends and fittings, and apply flux or CPVC primer as needed.

➤ Solvent-weld all plastic joints, or solder all copper joints, respectively.

➤ Be sure that the sillcock hangs at a slight downward slope to aid in draining.

➤ Secure the sillcock to the outside of your house.

Plumbing Perils

A frost-proof sillcock doesn't mean that you can leave your hose attached to it during freezing weather. The water in the hose can freeze and cause damage to the sillcock, which you won't discover until the water is next used, usually in the spring. Then the sillcock's pipe can start leaking as soon as it fills with water.

If for some reason you don't have a shutoff valve to your sillcock, now is the time to install one. A sillcock can be installed wherever you can conveniently tap into an existing cold water supply pipe.

Speaking of Freezing ...

Not only can your pipes freeze, but your water meter can, too. Prevention in this case is not only worth a pound of cure, but it also means that you can go without an expensive plumbing repair bill in the dead of winter. Regardless of how efficiently your plumber might work, everyone slows down when scrambling around unheated crawl spaces while dealing with frozen pipes.

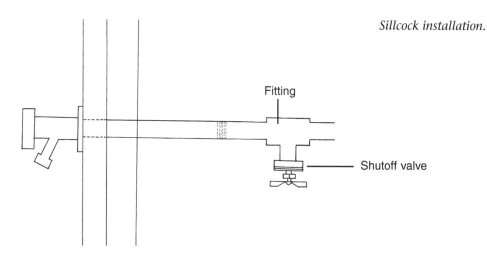

Sillcock installation.

Fitting

Shutoff valve

Water meters and main water supply pipes are usually located in basements and crawl spaces, the latter being unheated. Many basements are pretty utilitarian and are also unheated. To help prevent your meter or pipes from freezing during particularly cold weather, consider the following steps:

➤ Repair any broken widow glass.

➤ Consider installing storm windows and/or caulk around all windows.

➤ Insulate the walls.

➤ Purchase and install appropriate meter and pipe insulation from your plumbing supplier.

Plumbing the Depths

Besides freezing pipes, you could find yourself with a leaking main water supply pipe. Any leaks in this pipe are your problem—that is, any part of the pipe between the curb stop and your house. Any leaks between the curb stop and the city main are the utility's problem. If you notice any leaks outdoors (soggy ground in dry weather is a sure sign), attend to them right away.

If your pipes or meter do freeze, leave your propane torch alone! This isn't a time to use an open flame. Instead, use a hair dryer or a heat lamp to do the warming and thawing. Open the closest faucet to your meter or service line to allow any vapor from melting ice to escape. If you simply cannot thaw your pipes or meter, call a plumber.

Sprinkler Systems

The most basic sprinkler system is you watering your garden with a sprinkling can. This is fine if your garden consists of a few potted plants on the balcony of your Manhattan co-op, but even with a garden hose, hand watering gets a little old if you have the big house in the suburbs. Moving around an individual sprinkler works, but it isn't a very scientific or accurate way of watering. Usually we move them around when we think about them, meaning that some parts of the lawn get flooded and others might get watered less than an optimum amount.

A built-in sprinkler system and/or drip irrigation lines will make watering less of a chore and, in fact, will automate it. A timer allows you to set it and forget it, a real bonus when you're away on summer vacation and need to keep your prize roses happy. Installing a system means planning the locations of the sprinkler heads and pipe, digging trenches to lay the pipe, and connecting to the water supply. Before you begin, you'll need to look at your water supply.

Enough Water to Go Around

Water capacity is everything, whether you're adding a bathroom or fountains to rival those at the Bellagio Casino in Las Vegas (their fountains are a big attraction). You want to be sure that your water system can handle all the sprinkler heads that you want to install. System capacity is determined by these factors:

1. The size of your water meter (usually ⅝, ¾, or 1 inch)

2. The size of your service line

3. Your water pressure

A water meter's size is often stamped on the side of the meter, or it might be printed on your water bill. A service line may be smaller than the water meter size and therefore not be taking full advantage of the meter. Every sprinkler system is different and will state minimum system requirements, depending on the number of sprinkler heads and the length of pipe needed to irrigate your yard.

Planning Your System

Each sprinkler system comes with recommendations from the manufacturer for setting it up. You'll need to map out your yard, noting the locations of sidewalks, the driveway, and any other area that is not to be watered. Measure carefully. You don't want to be locating sprinklers in such a way that they douse your house as much as they do the yard or waste water by showering your driveway.

With your map of your yard, you can determine where water is needed, the most appropriate spray pattern, and the best type of sprinkler head to water the area. You want the spray to overlap so that the entire yard is covered. Most systems are broken down into zones, each controlled by a single valve, that supply water to sections of your yard with similar watering needs. This way, one set of sprinkler heads and their pipes can efficiently supply water on the same schedule. For instance, a sunny, exposed part of your yard will have different watering needs than one that is mostly shaded.

The Components

Each sprinkler system comes with some variation on the following parts:

➤ Pipe

➤ Valves

➤ Sprinkler heads

➤ An automatic timer

Old systems used galvanized or even copper pipe and brass sprinkler heads, but modern systems usually use PVC pipe to connect the service line to the sprinkler system's control valves. Polyvinyl chloride (PVC), flexible polybutylene (PB), or polyethylene (poly) can be used from the control valves to the sprinklers. PB pipe is not used to connect at the service line because it isn't strong enough to endure pressure surges. Poly is used in areas of severe winters or very rocky soil where PVC would not be appropriate. Plastic pipe is all solvent-welded, a much easier installation than copper or galvanized pipe.

Flexible pipe (PB and poly) are connected with special fittings (it was largely due to these fittings that gave PB its reputation for unreliability). Poly uses *insert fittings* that slide inside the two sections of pipe and then depend on a series of barbs to secure

Pipe Dreams

At least two manufacturers, Rain Bird and Lawn Genie, offer a free sprinkler system design service. Simply download one of the layout forms, fill it in with your property measurements according to the enclosed instructions, and return it to the respective manufacturer. They'll provide you with a finished plan and a list of needed components. Go to www.lawngenie.com/design.htm or www.rainbird.com/consumer/free/index.htm for this service.

Plumbing Perils

You don't want a sprinkler to be watering the side of your house or garage, especially if they're covered with wood siding. This can cause rot at worst and destroy your paint at best. Adjust your sprinklers so that they're soaking only your plants and lawn.

the pipes. These may work in theory, but you should use a *stainless steel clamp* to secure the fitting. Regardless of which pipe you select, purchase a strong grade so that you'll avoid future maintenance hassles. Cheap materials end up costing the most in the end. The last thing you want to have to do is dig up your sprinkler system to repair burst pipes.

What's That Thingamajig?

An **insert fitting** depends on prongs or barbs to dig into the inner surface of two flexible pipe sections. A **stainless steel clamp** wraps around and tightens the outside of an insert fitting to strengthen the connection.

Pipe Dreams

Dig your valve manifold deep enough to include several inches of gravel underneath the valves. The gravel makes an absorbent bed for any leaking water as well as drained water from the pipes. Install your manifold away from frequent foot traffic and close to the main water supply line. You may want a manifold in both the front yard and the backyard.

Your local plumbing code will outline which grades and types of pipe you can use for your sprinkler system. The pipe running from your water supply to the valves (the mainline), for instance, is always under pressure and for that reason must be heavier duty than the pipe used for the rest of the system.

Valves

Your system will have three kinds of valves, each with a separate purpose:

➤ A shutoff valve

➤ Control valves

➤ Drainage valves

A shutoff valve will run between your main water supply line and your control valve(s) and will allow you to turn off the sprinkler system. A shutoff gate or a ball valve is usually used for this purpose. A control valve supplies water to a section of sprinkler pipes and heads. These can be either manual valves or valves connected to a timer by a low-voltage wire. Control valves are installed either above ground or below in a valve manifold box. One control valve is used per irrigation zone.

Drainage valves are installed to drain the system at the end of the watering season. These valves come in two flavors: manual and automatic.

You must open a manual valve, well, manually, while automatic valves drain every time there's a drop in the pipes' water pressure. Some feel that this is a waste of water because it drains the system every time your sprinklers go through a cycle. It also means that the pipes must get refilled with pressurized water every time you water the yard. This frequent expansion and contraction against pipe connections can be a little tough on the system.

You will need a drain valve at each low point in the piping (the sprinklers at the high points will allow air in on the other end). Install the pipes so that their low points are just after the remote control valves and then slope upward to the sprinkler heads. This way, the drainage valves and control valves are all housed in one place inside the valve manifold.

Most local codes require that your valves have built-in backflow prevention devices. These prevent sprinkler water, which might get contaminated with yard dirt, from flowing back into your home water supply.

Heads Up!

Sprinkler heads come in a variety of styles. Each one serves a different watering purpose:

➤ Pop-ups

➤ Bubblers

➤ Shrub heads

Pop-ups are chiefly used for watering grass, while bubblers and shrub heads take care of flower beds, planters, and shrub areas. Your system's instructions will show you the best use and location of each type of sprinkler head. Modern heads are designed to keep out dirt, conserve water, and keep maintenance at a minimum.

Check with your code to see if antisiphon sprinkler heads are required (they're a good idea even if your code doesn't require them).

Plumbing the Depths

There's more to sprinklers than meets the eye. California State Polytechnic University, in Pomona, California, offers an entire program in irrigation with courses on such topics as micro irrigation, principles of irrigation, landscape hydraulics, landscape sprinkler irrigation (I and II), landscape drainage, automatic irrigation system controls, and landscape irrigation water management. So much for rolling out the garden hose and watering the lawn.

Time Out

Automatic timers are a great piece of technology. Set them and forget about ever watering your yard again. A timer does the following:

➤ Is programmed to run each control valve separately, according to the watering needs of its specific zone

➤ Saves water by running the system for only a set amount of time

➤ Allows for customization of your watering schedule

A timer must be connected to a power source. It uses a transformer to step down or reduce the standard 120-volt current to a lower, usable voltage. Your system's instructions will state the power requirements for the timer.

Pipe Dreams

You'll have to drill a hole through the wall when connecting to the main water supply pipe in the basement. In freezing climates, slope the pipe downward from the control valves into the basement. Use a shutoff valve with a drain cap so that you can drain off any water into a bucket at the end of the watering season. The pipe hole must be caulked, especially if it's drilled through the foundation. Foundation holes can really leak in wet climates. We found this out the hard way with our own house after the builder failed to caulk around the main water supply pipe.

Installing Your System

The hard part about installing sprinklers is the digging. The easy part is that you'll be using plastic pipe. Your first step, after you've planned your sprinkler locations and pipe runs, and purchased your materials, is to cut into your main water supply line and install a tee fitting for the system's shutoff valve. There's little point in digging up the water supply line outside and installing the valve when you can do it from inside your basement or crawlspace. It's a lot easier to do it inside your house.

Once the tee is in, a section of pipe will run from the tee to the shutoff (a ball valve or a gate valve). A second section of pipe will then run from the shutoff to the control valve(s), which may be buried outside in a valve manifold or mounted above the ground. Once this pipe is run to the new location of the control valves, but before the valves are installed, you can start digging.

Digging Fools

In warm climates, you don't have to worry about freezing pipes—well, not usually, but even Florida gets a freeze once in a while. Check your local code for depth requirements when burying your sprinkler pipes. Note that the mainline, the section between the shutoff valve and the control valves, might have to be deeper than the other pipes. Commercial irrigation systems must be buried at least 12 inches deep. Before doing any digging, check with your local utilities for the location of water pipes, gas lines, and electrical wires.

Your system should use shorter, direct branch lines rather than long runs with multiple turns to conserve water pressure. Pipes and fittings should be the same size as the control valves (three-quarter-inch, for instance). Lay out your system by pounding wood stakes into the ground at each sprinkler head and control valve location. Connect the stakes with string or bright yellow construction tape, and check your layout. When you're satisfied with its design, you begin digging the trenches either by hand or with a rented trencher. A trencher can be unwieldy and is usable only on flat lawns or empty garden beds.

If you're digging by hand, keep these tips in mind:

➤ Soak the yard a day or two before hand to soften the soil.

➤ Carefully cut away about 8 to 10 inches of sod with a spade (a flat shovel), and set it aside, roots down, on a narrow sheet of plastic. Keep the sod wet until you replant it.

➤ Dig your trench as deep as your local code requires.

If you have to leave the trenches exposed overnight, be sure to cover them with planks so that no one gets injured by falling in.

Pipe Dreams

If you have to go under a sidewalk, attach a piece of pipe that has a hose-pipe adapter on one end to your garden hose. Poke the other end of the pipe into the dirt where you have to tunnel, and turn on the water. The water coming out that end will wash away the dirt, forming a cavity for inserting the sprinkler pipe.

Installing the Control Valves

If possible, go ahead and install your control valves above ground. Otherwise, construct a valve manifold and install them underground. Fittings and installations will vary depending on the system you use. Be sure that your installation follows local code requirements for antisiphon valve heights.

Lay out the control valves and connectors, and install them according to your system's directions and requirements. Allow the solvent to cure, turn off all the control valves, and turn on the water at your shutoff valve. Check for leaks and proceed with the pipe installation.

Hooking Up

Lay your first run of pipe from its control valve to the first sprinkler head. A tee will be inserted into the pipe at every sprinkler head, except for the last on at the end of the run; this will get a 90° elbow. Every tee and the final elbow will have a riser, or vertical pipe, attached to it to hold the sprinkler head. The risers must be cut to an appropriate height to accommodate the sprinkler head (refer to your system's instructions).

After the risers are installed, you'll run another test for leaks and flush the system at the same time.

➤ Install a pipe plug in every riser except the last one on each zone or pipe.

➤ Turn on the water at the shutoff valve, and open the control valves one at a time.

➤ Check the entire system for leaks.

➤ Shut off the water at the control valves, and remove the plugs.

Aside from testing for leaks, this also will flush out any dirt that may have gotten into the pipes during your installation.

Plumbing Perils

If you don't want to tackle the entire installation of your sprinkler system, consider doing the trenching, following your installer's plan. Compare costs first, however. An installer might own a trenching machine and do the digging pretty economically.

Installing the Sprinkler Heads

Again, the installation depends on the type of head and your manufacturer's instructions. If you use Teflon tape on the threads, be sure it's installed neatly so you'll get a tight connection. Pop-up sprinkler heads should be installed only slightly above the *soil* surface, not the grass. Otherwise, they'll get damaged by lawnmowers. Shrub heads and bubblers should be mounted several inches above the soil surface for a maximum watering radius.

Time to Water

Normally, the best time to water is early in the morning before the temperature heats up too much. This way, you lose less water through evaporation and the water can really soak in. Unlike running your sprinklers late at night, early morning watering won't leave your plants damp all night and give them an opportunity to develop mildew or rotted roots.

Winter's Coming

Even if your system's water supply pipes are buried below the frost line, the vertical tubing or pipe connected to the sprinkler heads is not. In a cold climate, you'll want to drain the system before freezing weather shows up. The easiest way is to install either automatic or manual drain valves at the low points in the system's pipes. Otherwise, the system will have to be blown out with compressed air.

Plumbing the Depths

Some people really want to spread the word about sprinkler and irrigation systems. Jess Stryker has an entire Web site devoted to sprinkler systems and their installation. Go to www.JessStryker.com/spklr.htm and learn more about sprinklers than you thought possible.

To drain and prepare your system for winter, follow these tips:

➤ Shut off the water supply to the system.

➤ If your system has an automatic timer, shut it down according to the manufacturer's instructions.

➤ Open the system's drain valves.

➤ If your manufacturer recommends it, consider having your system blown out by a sprinkler technician.

Plastic pipe must be treated gingerly when compressed air is being shot into it. One mistake with a compressor, and you can split a pipe or launch a sprinkler head.

Wrap any exposed pipes with pipe insulation. Pack the valve manifold with fiberglass insulation stuffed inside plastic bags to protect the valves from the cold. And follow any additional recommendations from your system's manufacturer.

Plumbing Perils

One way to avoid leaks caused by cold temperature in your system is to bury all the pipes below the frost line, even if your code doesn't require it. Drain valves installed at the system's low point will ensure that the pipes are empty at the end of the watering season.

Spring Is Just Around the Corner

Just as you have a winter shutdown, you also have a spring startup with your system. You can't just slam open the valves and start watering (well, you can, but it just isn't a good idea). A sudden jolt of water could possibly cause some damage.

Your first step is to remove any insulation from your valve manifold. Next, open the shutoff valve *slowly*. Open the control valves one at a time and check for leaks. Sometimes their seals dry out over the winter and need to be replaced.

Check your timer, and be sure that each station is properly programmed. Replace any backup battery with a new alkaline battery. Review your system's instruction manual for other specific startup procedures.

The Least You Need to Know

➤ A frost-free sillcock is a must in cold climates.

➤ Plan your sprinkler system carefully before you begin installing it.

➤ Like sillcocks, sprinkler systems need to be drained if you live in an area of freezing winter temperatures.

➤ Modern sprinkler systems are automated and set up by zones, making your watering chores almost effortless.

Part 6

Other Plumbing Concerns

You didn't think you were finished already, did you? There's always more plumbing that you can add or improve upon. One major plumbing appliance that could use some routine maintenance, if not actual replacement, is your hot-water heater. None of us wants to go without hot water, and a little attention to your water heater will reap benefits that you can enjoy every time you step into the shower.

Although I don't recommend that you replace your hot-water heater yourself, it helps to understand the mechanics of the tank and to recognize the signs that call for its replacement; we'll cover these in Chapter 23, "Hot Water Heaters." Laundry rooms also have plumbing requirements that can be provided in several ways; Chapter 24, "Laundry Rooms," gives you a clearer idea of how to keep your washing machine—and you—happy.

Although rarely found in new housing, hot-water and steam-heating systems were used more often through the 1930s, especially in big homes and apartment buildings. If you ever wondered how these clanky, strange systems worked, we'll go over them in Chapter 25, "Heat and Plumbing Concerns."

Chapter 23

Hot Water Heaters

In This Chapter

➤ How water heaters work

➤ Choosing a water heater

➤ Common problems

➤ Alternative systems

After toilets, it's safe to say that our next favored plumbing convenience is a hot water heater. Hot water brings relaxation while soaking in a tub and helps sanitize when we clean. Bathing was done far less frequently in earlier times, in part because water had to be laboriously boiled on stoves and poured into a tub. The other reasons had to do with peculiar religious and health beliefs, but those are another matter.

The most popular style of hot water heater is a storage tank that takes in cold water, retains it, heats it to a preset temperature, and stands ready to distribute hot water when a tap is opened. You can choose among gas, oil, electric, or solar water heaters and different sizes of tanks. The best size of water heater for your home is determined by how many people it will be serving during peak demand periods. A family of six will require more hot water than a single person (they should anyway). An undersized water heater won't deliver sufficient quantities of hot water to meet demand.

The optimal water heater size is also affected by its recovery rate. This is the water heater's ability to heat incoming cold water to a preset temperature. The recovery rate is determined by the size of the heating unit and the type of fuel used.

A hot water heater doesn't require much care, but its life can be extended with regular maintenance. Its temperature must be set to avoid scalding, especially in the shower. A tank also must be strapped to the wall to prevent it from tipping over in the event of seismic activity. This chapter covers the operation, installation, and repair of hot water heaters.

How Water Heaters Work

Water heaters supply hot water to fixtures and appliances. As hot water is drawn out from the top of the tank, cold water replaces it through a dip tube that empties near the bottom of the tank. The heater's thermostat maintains a preset temperature by sensing any drop in the water's temperature and activating the gas or oil burner or electric elements to heat the water further. If the thermostat fails and the water overheats dangerously, a temperature-pressure relief valve opens, allowing hot water to flow through a discharge pipe on the outside of the tank. This relieves pressure inside the water heater and prevents an explosion.

At the base of a water heater is a drain valve that allows you to drain water for periodic inspection for sediment buildup and to empty the tank. A water heater has other components as well:

➤ An anode rod

➤ Insulation

➤ A shutoff valve

Plumbing Perils

Steaming hot water coming out of your faucet is a sign that your water heater is overheating, probably the result of a faulty thermostat. Turn off the power source (gas or electricity) and open the hot water taps until the water runs cold. Turn off the faucets and call a plumber to inspect and repair the tank.

The anode rod, which is often made from magnesium, hangs inside the tank and protects its lining from corrosion. The anode or sacrificial rod attracts ions in the water that would otherwise oxidize the steel lining. It's a lot easier and cheaper to replace an anode rod than the entire tank.

Insulation is installed in the tank walls to help maintain water temperature. The amount of insulation determines the water heater's efficiency rating as measured by the American Society of Heating, Refrigerating, and Air Conditioning Engineers (ASHRAE). A rating of 90, for instance, tells you that the water heater is well-insulated and does not require an exterior insulation blanket. If your water heater has a lower rating, ask your energy provider about an insulation blanket and its cost.

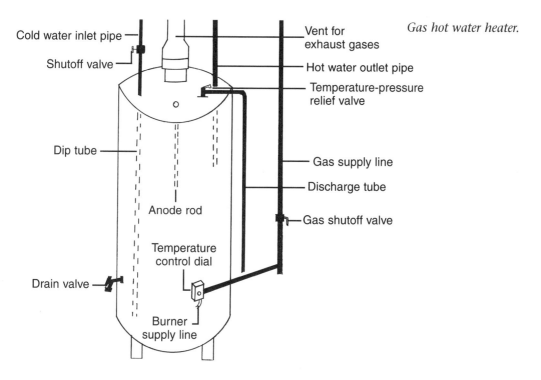

Gas hot water heater.

Cold water inlet pipe

Shutoff valve

Dip tube

Anode rod

Temperature control dial

Drain valve

Burner supply line

Vent for exhaust gases

Hot water outlet pipe

Temperature-pressure relief valve

Gas supply line

Discharge tube

Gas shutoff valve

Decisions, Decisions

You may be thinking of installing a new hot water heater and possibly changing to a different energy source. Generally, oil and gas water heaters have a faster recovery rate over electric water heaters, but some electric models have fast recovery, dual-element tanks that recover faster than standard models. You have to consider these factors:

➤ The size of the tank (typically 30, 40, or 50 gallons, although larger sizes are available)

➤ How much hot water you require

➤ How hot you want the water to be heated

➤ Whether your pipes and the water heater are insulated

Generally speaking, it's cheaper to heat with natural gas. Electricity is a more efficient energy source at the point of use, but gas is more efficient (and cheaper) at the point of production and distribution. Electricity incurs losses in efficiency as it's produced from fossil fuels (you're burning one energy source to create another one) and as it's transmitted along power lines. Natural gas, on the other hand, loses some of its efficiency as a fuel when it heats the water because it loses heat up the flue as byproducts of combustion are carried off. Before installing a new water heater and changing to a new energy source, be sure to determine the following:

➤ Your local energy costs

➤ The cost of conversion

➤ The difference in the cost of the water heaters themselves and their recovery times

➤ The payback period

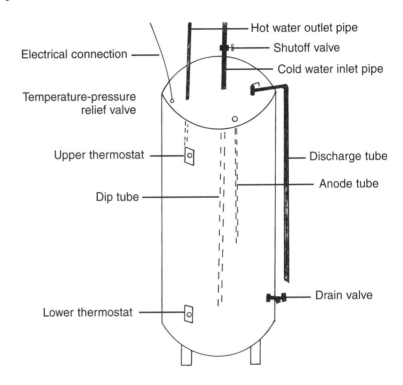

Electric hot water heater.

Electrical connection

Hot water outlet pipe

Shutoff valve

Cold water inlet pipe

Temperature-pressure relief valve

Upper thermostat

Discharge tube

Anode tube

Dip tube

Lower thermostat

Drain valve

Plumbing the Depths

Energy providers compete for customer accounts and dollars. Your local gas company might be willing to run a gas line to your house at no cost or nominal cost when you switch to a gas hot water heater or furnace. Other providers offer their own incentives to retain your account or encourage you to switch. Factor in all the costs before making a decision.

One advantage to an electric water heater is the absence of any combustible fuel and the need to install a flue. On the other hand, if your power goes off for an extended period of time, you'll not only be living in the dark, but you'll be taking cold showers as well. If you live in an area of regular power outages, this may be a determining factor.

Reading the EnergyGuide Label

Hot water heaters carry an EnergyGuide Label that will state its energy efficiency ratings or its annual estimated operating cost. It also states the peak hour demand capacity or first-hour rating (FHR). This tells you how much hot water the system can pump out during times of the highest demand. You'll want a hot water heater that can match the peak hour demand in your home. This is often the morning hours when bathrooms and kitchens are getting heavy use from showering, washing, and cleaning dishes. Gas models have higher FHRs than electric models, so you might be able to install a smaller gas hot water heater and still meet your hot water needs.

Some EnergyGuide labels also state the water heater's energy efficiency rating, or EF (the higher the rating the better). If the label doesn't state it, the manufacturer's literature should.

Plumbing Perils

New water heaters come with insulation sandwiched between the outside of the tank and the inner lining. Old tanks can be made more efficient with an insulation blanket wrapped around the outside. Although you can wrap the sides and top of an electric water heater, you should never insulate the top or bottom of a gas model.

Cost Comparisons

The initial price of a hot water tank is only part of your overall cost. You should total up all the costs, including these:

➤ Installation

➤ Price of operation (energy cost)

➤ Maintenance over the system's lifetime

A water heater that is inexpensive to purchase could be the most expensive to maintain. On the other hand, the unit with the longest warranty will cost more to purchase but less to maintain. Features to look for include these:

➤ Self-cleaning tank (reduces sediment)

➤ Stainless steel elements

➤ Energy-conserving heat traps

How much will all this cost? Not much if a recent advertisement in our local paper by a national chain of home improvement stores is any indication. The stores were offering a 50-gallon General Electric hot water heater with a 12-year limited warranty for $298. By the end of the warranty period, it will cost you a little more than $2 a month to purchase (installation would be extra), roughly the price of an espresso.

Some utility companies might lease you a water heater at a competitive rate that includes maintenance. If that's the case, you can skip this chapter entirely!

Hot Water Problems

Newer hot water heaters are pretty reliable. By the time you need to replace valves, for instance, you'll probably have to replace the tank. Both new and older units can experience nuisance problems that are easily solved:

➤ No hot water

➤ Water too hot

➤ Not enough hot water

A lack of hot water in a gas hot water heater suggests that the pilot light has gone out and needs to be lit; in an electric unit, the fuse might have burned out or a circuit breaker might have been tripped. These are the immediate causes to look for. Water that is too hot can simply be a matter of adjusting the temperature control. Insufficient hot water might be due to excessive use or a too-low temperature setting. Sometimes these symptoms can point to larger problems.

Pipe Dreams

One way to keep your hot water heating costs down is to lower the temperature when you're on vacation or away from the house for more than 48 hours. Modern tanks have a vacation setting on the temperature control dial for this purpose.

Pipe Dreams

If you're running out of hot water too quickly, check your shower head(s). Replacing an old shower head with a new restricted–flow model can cut its hot water outflow in half. This is an inexpensive way of altering your hot water situation.

This Water's Cold!

Lukewarm or cool water can result from an internal problem with the tank. A broken or worn dip tube (more later) prevents incoming cold water from being properly heated by an element in the bottom of the tank. In electric water heaters, one of the two heating elements can be malfunctioning (check the Reset button, which might need to be pushed or reset). In gas water heaters, the thermocouple or burner could be malfunctioning.

The Dip Tube Capers

A dip tube empties cold water into the bottom of a hot water heater so that it will get heated and rise toward the top of the tank as hot water is drawn off. If the tube were to empty cold water near the top of the tank, you would end up with lukewarm water. Problems with polypropylene dip tubes manufactured between August 1993 and February 1997 by the major dip tube supplier in the country were the subject of a class action settlement by major hot water heater manufacturers.

The disintegrating tubes have clogged faucets, dishwashers, and shower heads, and they affect hot water temperatures by cooling the water. If your hot water heater was manufactured in the period from 1993 to 1997 period, and if you're experiencing problems with lukewarm water and have had to raise the temperature control, you could be entitled to a free warranty repair. One additional symptom to watch for is the presence of small white particles, which are bits of the deteriorating dip tube, in your hot water. Contact the manufacturer of your tank for more information, or go to www.diptube.com. Dip tube replacement is usually best left to a plumber.

Temperature-Pressure Relief Valve

Hot water heaters have a built-in safety system with the temperature-pressure relief valve. It prevents excess pressure from building up inside the tank by opening and releasing excessively hot water or steam out the discharge pipe. You should test this valve every few months by following these steps:

➤ Place a bucket under the discharge pipe.

➤ Lift the lever of the valve to release the water.

➤ Repeat and lift the lever a few more times to clear out sediment.

Plumbing the Depths

Since the invention of the steam engine, boilers have been blowing up on a regular basis. In 1997, there were 2,455 boiler accidents, resulting in 18 deaths and 75 injuries, according to the National Board of Boiler and Pressure Vessel Inspectors. The board states that a 30-gallon home water heater, which is a type of boiler, has enough energy to lift the average family car 14 stories traveling at 85 MPH, representing more than two million foot-pounds of energy. Pressure-temperature valves have been installed on boilers since the late nineteenth century, but accidents still occur.

If no water flows out the valve, or if water continues to drip out of the discharge pipe, the valve should be replaced. Some relief valves empty into a small storage tank instead of emptying through a discharge pipe. It's unlikely that you'll ever need to replace a relief valve, but if you do, it's another good job for a plumber.

The Tank Isn't Leaking!

In humid climates, condensation can appear on the outside of your water heater if a large outflow of hot water occurs and is replaced with cold water. This can look like your tank is leaking at the bottom. To check, wipe down the tank completely and observe how long it takes for the drips to reappear. If they appear immediately, you may well have a tank leak, in which case your hot water heater will have to be replaced. Otherwise, it's just normal condensation.

Sediment

Sediment buildup in the bottom of a hot water heater results from the heat precipitating calcium carbonate out of the water. The sediment can cover the heating element or burner, causing them both to work much harder to heat the tank's water. Once the sediment has built up to this degree, your only chance of getting rid of it is to drain the tank. To drain a hot water heater, do this:

➤ Disconnect the fuse or shut off the circuit breaker to an electric model; with a gas model, turn the temperature control to off and close the gas shutoff valve.

➤ Turn off the cold water at the cold water shutoff valve.

➤ Attach a hose to the drain valve, and run it to a floor drain (the hose must be lower than the drain valve) or into a bucket.

➤ Open several hot water valves in your house.

➤ Open the drain valve and drain all the water from the tank.

➤ Run a few gallons of cold water into the tank, and let it drain (this will pick up additional sediment).

➤ When the tank has emptied, close the drain valve, disconnect the hose, and open the cold water supply shutoff valve.

➤ The tank will be full when water flows from the faucet farthest from the tank.

➤ Close any open faucets, and turn on the power or light the pilot light (instructions will be on the outside of the water heater).

Plumbing Perils

Be careful when draining a hot water tank. If the water is too hot, you can get scalded if it spills on your hands. Fill your bucket about three-quarters full with the hot water, and then empty it. Better yet, use a hose, if possible, and run the water into a drain.

To avoid sediment buildup, drain a gallon or two of water from the tank every month or so.

You Hear a Noise?

Another problem associated with sediment buildup in gas water heaters is noise. As sediment settles at the bottom of the tank, gas bubbles from the burners form underneath it. As they try to escape and rise toward the top of the tank, the bubbles form a popping sound, which can be a little disconcerting. Cleaning out the sediment will end this problem, too.

Watch That Temperature!

Setting your hot water heater temperature to 120° helps prevent scalding, particularly with young children and the elderly. People in these age groups are more apt to get burned from hot water because of their slower reaction times, especially in tubs and showers. It doesn't take much exposure to become seriously burned. According to the Shriners Burn Institute, scalding occurs in about 30 seconds when the water temperature is 130° and less than five seconds at 140°. Some municipalities mandate a 120° setting when a hot water tank is installed, but they cannot really control any subsequent temperature adjustment. It's up to the homeowner to keep the temperature lower.

Securing Your Tan

You want your water tank to remain upright and secure in the event of an earthquake. You might think that this would be a concern mainly for Californians, but according to the National Earthquake Information Center (NEIC), every one of the 50 states has had some earthquake activity at some point in its history. Sure, Connecticut had its last big one in 1791, but you never know when history will repeat itself.

Hot water heaters are heavy. A 50-gallon tank contains close to 400 pounds of water. If it were to topple over, it would damage itself, its electrical or gas connections, its cold water connection, and anything in its path. To prevent this kind of damage, a hot water tank should be strapped to a nearby wall or floor with *plumber's tape.*

The Los Angeles Fire Department recommends the following steps to secure your hot water heater:

Pipe Dreams

You sometimes can find wall studs by knocking across the wall with your knuckle and listening for a change in the sound—the wall will sound more solid when you knock across a stud. Otherwise, buy an inexpensive battery-powered stud finder at a hardware store. These tools work on drywall, but are not effective on plaster walls. Studs typically are installed 16 inches on center in new houses.

What's That Thingamajig?

Plumber's tape is a perforated, flexible metal strapping used to secure pipe to framing lumber, usually overhead floor joist. It is also used to secure water heaters to prevent them from toppling over in the event of an earthquake.

➤ Mark your water heater 6 inches down from the top and 4 inches above the thermostat control valve at the bottom.

➤ Mark the studs on the nearby wall at the same points as those on the water heater, but go out at least 24 inches wider then the water heater.

➤ Secure the top of the tank by drilling a ⁵⁄₁₆-inch pilot hole into the farthest stud from the tank. Secure one end of the tape to this stud using a ⁵⁄₁₆-inch by 2½-inch lag bolt and washer.

➤ Secure the tape to the next-closest stud in the same manner. (You want to secure the tape to at least two studs on each side of the tank.)

Plumbing Perils

Your insurance company might require you to strap and secure your water heater as part of your homeowners policy. Be sure that your strapping method meets with their requirements. Otherwise, they may not be responsible for damages from a toppled tank.

➤ Wrap the tape completely around the tank in a loop, and secure the other end of it to two studs on its other side in the same manner as the first end of the tape was secured. The tape should be tight, with no slack.

➤ Repeat this procedure with the lower tank strap.

➤ Insert and secure a nonflammable spacer between the tank and the wall to prevent the water heater from moving toward the wall.

➤ Secure angle brackets to the floor, and secure the legs of the water heater to the angle brackets.

These guys are serious and their recommendations should be followed. It's a small job, given the potential damage if your water heater ever does fall over.

Alternatives to Tanks

A instantaneous or on demand hot water system is a tankless system. Instead of heating up a large volume of water and keeping it hot, whether there's any demand for it or not, a tankless unit heats water whenever there is a demand for it. Open up a faucet or start the shower and cold water is heated as it passes through the unit. This technology is available as a whole house system or at an individual point of use (a bathroom sink, for instance). Manufacturers of tankless systems seem to believe that they are the saviors of Western civilization while manufacturers of traditional hot water heaters believe otherwise. Once again, I'll let the government (in this case, the Department of Energy) offer its crisply written view.

Demand (Tankless) Water Heaters

Water heating accounts for 20% or more of an average household's annual energy expenditures. The yearly operating costs for conventional gas or electric storage tank water heaters average $200 or $450, respectively. Storage tank-type water heaters raise and maintain the water temperature to the temperature setting on the tank (usually between 120°-140°f (49°-60°c). The heater does this even if no hot water is drawn from the tank (and cold water enters the tank). This is due to "standby losses": the heat conducted and radiated from the walls of the tank-and in gas-fired water heaters-through the flue pipe. These standby losses represent 10% to 20% of a household's annual water heating costs. One way to reduce this expenditure is to use a demand (also called "tankless" or "instantaneous") water heater.

Demand water heaters are common in Japan and Europe. They began appearing in the United States about 25 years ago. Unlike "conventional" tank water heaters, tankless water heaters heat water only as it is used, or on demand. A tankless unit has a heating device that is activated by the flow of water when a hot water valve is opened. Once activated, the heater delivers a constant supply of hot water. The output of the heater, however, limits the rate of the heated water flow.

Gas and Electric Demand Water Heaters

Demand water heaters are available in propane (lp), natural gas, or electric models. They come in a variety of sizes for different applications, such as a whole-house water heater, a hot water source for a remote bathroom or hot tub, or as a boiler to provide hot water for a home heating system. They can also be used as a booster for dishwashers, washing machines, and a solar or wood-fired domestic hot water system.

You may install a demand water heater centrally or at the point of use, depending on the amount of hot water required. For example, you can use a small electric unit as a booster for a remote bathroom or laundry. These are usually installed in a closet or underneath a sink. The largest gas units, which may provide all the hot water needs of a household, are installed centrally. Gas-fired models have a higher hot water output than electric models. As with many tank water heaters, even the largest whole house tankless gas models cannot supply enough hot water for simultaneous, multiple uses of hot water (such as, showers and laundry). Large users of hot water, such as the clothes washer and dishwasher, need to be operated separately. alternatively, separate demand water heaters can be installed to meet individual hot water loads, or two or more water heaters can be connected in parallel for simultaneous demands for hot water. Some manufacturers of tankless heaters claim that their product can match the performance of any 40 gallon (151 liter) tank heater.

Selecting a Demand Water Heater

Select a demand water heater based on the maximum amount of hot water to meet your peak demand. Use the following assumptions on water flow for various appliances to find the size of unit that is right for your purposes:

➤ Faucets: 0.75 gallons (2.84 liters) to 2.5 gallons (9.46 liters) per minute.

➤ Low-flow showerheads: 1.2 gallons (4.54 liters) to 2 gallons (7.57 liters) per minute.

➤ Older standard shower heads: 2.5 gallons (9.46 liters) to 3.5 gallons (13.25 liters) per minute.

➤ Clothes washers and dishwashers: 1 gallon (3.79 liters) to 2 gallons (7.57 liters) per minute.

Unless you know otherwise, assume that the incoming potable water temperature is 50°f (10°c). You will want your water heated to 120°f (49°c) for most uses, or 140°f (60°c) for dishwashers without internal heaters. To determine how much of a temperature rise you need, subtract the incoming water temperature from the desired output temperature. In this example, the needed rise is 70°f (21°c).

List the number of hot water devices you expect to have open at any one time, and add up their flow rates. This is the desired flow rate for the demand water heater. Select a manufacturer that makes such a unit. Most demand water heaters are rated for a variety of inlet water temperatures. Choose the model of water heater that is closest to your needs.

As an example, assume the following conditions: one hot water faucet open with a flow rate of 0.75 gallons (2.84 liters) per minute. One person bathing using a shower head with a flow rate of 2.5 gallons (9.46 liters) per minute. Add the two flow rates together. If the inlet water temperature is 50°f (10°c), the needed flow rate through the demand water heater would need to be no greater than 3.25 gallons (12.3 liters) per minute. Faster flow rates or cooler inlet temperatures will reduce the water temperature at the most distant faucet. Using low-flow showerheads and water-conserving faucets are a good idea with demand water heaters.

Some types of tankless water heaters are thermostatically controlled. They can vary their output temperature according to the water flow rate and the inlet water temperature. This is useful when using a solar water heater for preheating the inlet water. If, using the above example, you connect this same unit to the outlet of a solar system, it only has to raise the water temperature a few degrees more, if at all, depending on the amount of solar gain that day.

Cost

Demand water heaters cost more than conventional storage tank-type units. Small point-of-use heaters that deliver 1 gallon (3.8 liters) to 2 gallons (7.6 liters) per minute sell for about $200. Larger gas-fired tankless units that deliver 3 gallons (11.4 liters) to 5 gallons (19 liters) per minute cost $550-$1000.

The appeal of demand water heaters is not only the elimination of the standby losses and the resulting lower operating costs, but also the fact that the heater delivers hot water continuously. Gas models with a standing (constantly burning) pilot light, however, offset the savings achieved by the elimination of standby losses with the energy consumed by the pilot light. Moreover, much of the heat produced by the pilot light of a tank-type water heater heats the water in the tank; most of this heat is not used productively in a demand water heater. The exact cost of operating the pilot light will depend on the design of the heater and price of gas, but could range from $12 to $20 per year. Ask the manufacturer of the unit how much gas the pilot light uses for the models you consider. It is a common practice in Europe to turn off the pilot light when the unit is not in use.

An alternative to the standing pilot light is an intermittent ignition device (IID). This resembles the spark ignition device on some gas kitchen ranges and ovens. Not all demand water heaters have this electrical device. You should check with the manufacturer for models that have this feature.

Life Expectancy

Most tankless models have a life expectancy of more than 20 years. In contrast, storage tank water heaters last 10 to 15 years. Most tankless models have easily replaceable parts that can extend their life by many years more.

(Bibliography available at www.eren.doe.gov/ consumerinfo/refbriefs/bc1.html)

EREC (Energy Efficiency and Renewable Energy Clearinghouse) is operated by NCI Information Systems, Inc. for the National Renewable Energy Laboratory/U.S. Department of Energy. The statements contained herein are based on information known to EREC at the time of printing. No recommendations or endorsement of any product or service is implied if mentioned by EREC.

Pipe Dreams

The Department of Energy (www.doe.gov) is a terrific source of information on solar heating as well as water conservation. As a government service, they'll send you more research and booklets than you ever imagined.

 P.O. Box 3048 Merrifield, Va 22116

 1-800-DOE-EREC

 E-mail: doe.erec@nciinc.com

Let the Sunshine In

As a Seattle-area resident who lives with overcast skies, I've heard rumors about this great source of heat and light called the sun that can be used to heat water. In some parts of the country, such as Florida, solar water heating is a more viable option. Nevertheless, I would do plenty of research before you start tearing out your water heater while telling your neighbors that you're no longer going to be a slave to the power companies once you install your roof top solar collectors. They may remember your gloating the first time you knock on their door to use their shower because your tap water is 2° above freezing.

Even in the climates most favorable to solar usage, you need a conventional hot water heater as a backup system. Local building codes often require a backup, and codes or covenant restrictions in your neighborhood might dictate restrictions on collector sizes and their locations.

There are passive solar water heaters and active heaters. The Department of Energy (DOE) describes them (www.doe.gov) in this way:

> Active systems use electric pumps, valves, and controllers to circulate water or other heat-transfer fluids through the collectors. They are usually more expensive than passive systems but are also more efficient. Active systems are usually easier to retrofit than passive systems because their storage tanks do not need to be installed above or close to the collectors. But because they use electricity, they will not function in a power outage. Active systems range in price from about $2,000 to $4,000 installed.
>
> Passive systems move household water or a heat-transfer fluid through the system without pumps. Passive systems have no electric components to break. This makes them generally more reliable, easier to maintain, and possibly longer lasting than active systems. Passive systems can be less expensive than active systems, but they can also be less efficient. Installed costs for passive systems range from about $1,000 to $3,000, depending on whether it is a simple batch heater or a sophisticated thermosiphon system.

Such a deal! On top of that, you have to size your water heater in accordance with your household requirements as coordinated with your rooftop collector area. These formulas vary depending on where you live (the sunny South or the colder North).

According to the U.S. Department of Energy, the economic benefits of a solar water heater compare favorably with those of electric water heaters, but that's not so with less expensive natural gas. The economics are situational and depend on the cost of the system, your local cost of conventional fuel or electricity, the real life of the system and its maintenance costs, and the percentage of your hot water demands that it can meet. Because solar water-heating systems are more expensive than conventional water heaters, you have to figure in what return you could be getting on the money if it weren't tied up in the solar system. And remember, even with a solar system you still need a backup conventional system, which is more than a little ironic, considering that the solar heater is supposed to be saving you money and freeing you from depending on conventional fuel or electricity.

Some utilities offer rebates when you install a solar water heater, and some local and state governments toss in some tax incentives to encourage installations.

Before you decide to heat your water with sunlight, tighten up your existing system. The same applies if you simply plan to upgrade to a larger tank because you're running out of hot water too quickly. Follow some simple conservation steps first:

➤ Install low-flow shower heads.

➤ If your hot water heater is an uninsulated model, install an insulating blanket.

➤ Set your water heater temperature at 120°.

➤ Insulate your exposed hot water pipes.

Try staggering your shower times for your tank to recover while it heats more water. These changes may be enough to allow you to keep your current water heater.

Solar systems aren't quite as straightforward as conventional hot water heaters. Look for certified systems, such as those labeled by the Solar Rating & Certification Corporation with its SRCC OG-300 label. Check the system's warranty, and confirm that the installers are qualified technicians. The American Solar Energy Society (ASES, 2400 Central Ave., Unit G-1, Boulder, CO 80301 or 303-443-3130) is a good place to start for basic solar information.

The Least You Need to Know

➤ Water heaters are very reliable, but they still need regular maintenance and inspecting.

➤ Many factors, including the size of your household and the power source, will affect your choice of water heaters.

➤ The pressure-temperature relief valve is your water heater's most important safety feature.

➤ Regularly draining some water from your heater will help prevent sediment buildup at the bottom of the tank.

➤ It's important to strap and secure your water heater to a wall to prevent it from toppling over during seismic activity.

Chapter 24

Laundry Rooms

In This Chapter

➤ The advantages of laundry rooms

➤ Washing machine features

➤ Efficient laundry practices

➤ Installing a washing machine

➤ Drain line requirements

➤ Comments on clothes dryers

Most of us had at least one period in our lives, usually in college or a first apartment, when we used shared laundry facilities. We kept plopping quarters into washing machines and dryers after cramming a week's worth of laundry inside, and no doubt paid for the landlord's purchase price over and over again. Part of being a full-fledge adult meant having your own laundry room so that you didn't have to share with anyone. For all you downtown urban dwellers, well, I guess you're not adults yet, but your day will come.

Washing machines and dryers often are relegated to the basement, but their locations vary depending on the house. They also end up in garages or in separate utility rooms off the kitchen in many one-story homes. Slowly, laundry facilities are finding their way to the second floor of two story houses, a logical location because that's where most of the dirty linen and clothes are also located.

From a plumbing standpoint, a laundry room needs hot and cold water supplied for the washing machine and the laundry sink as well as drains and vents for both. New homes come with a hot and cold hose bib and a drainpipe for the washing machine and the rough-in plumbing for a sink. You supply the sink and appliances, unless you've made other arrangements with the builder. Older homes typically have existing laundry facilities that might need updating, or you might need to move them to another area. And, of course, there are always newer appliances to buy!

This chapter covers the laundry room basics, including some ways to conserve space and water. You might look back on your laundromat days with a certain nostalgia, but there's no comparing that to the convenience of having your very own laundry room.

Plumbing Perils

You can relocate a basement laundry area to just about any part of the basement, but it might take a lot of effort and work, such as digging through the floor to access the drain line. Consider carefully whether the move will be worth the effort.

What's That Thingamajig?

Laundromat refers to a service mark for a self-service laundry. A launderette is a self-service laundry, but *laundromat* seems to have won out as the preferred term.

It Beats a Washboard

There's an argument that suggests the more technological conveniences we have, the more work we do because they avail us to so many more options than we had previously. People used to cook all their meals in one pot over the fire or roast fresh game. Silverware was a luxury. Now we have convection ovens, microwave ovens, toaster ovens, food processors, and battery-operated spaghetti forks. We can create and prepare the most elaborate meals possible, spending hours more preparation than our ancestors who threw everything in the pot for true pot luck.

Laundry is different. Who wants to walk around in rank-smelling, dirty clothes? No one does, and we also don't want to return to washing in the nearest stream, beating out the dirt with rocks. Washing machines and dryers are wonderful utilitarian appliances, especially when you consider the alternative. Laundry rooms, and *laundromats,* for that matter, have evolved into cleaner, brighter facilities with their own particular plumbing and electrical needs.

Out of the Dark Ages

In older homes, laundry areas usually are tucked away in an unfinished basement. This was really the only logical place for them when these houses were built given the prevailing designs at the time. Basements were usually meant for storage and utilities. No one considered whether Mom—and it was always Mom, unless there was hired help—would mind carrying clothes up and down another flight of stairs.

Regardless of where your laundry area is located, you can design it the same way you do a kitchen or bathroom: Use plenty of built-in cabinets, counters, and shelves. Consider what you do in a laundry room, and design it accordingly:

➤ Build in adjustable shelves for detergent and other laundry supplies.

➤ Add closet rods to hang drip-dry clothes or stay-press clothing as it comes out of the dryer.

➤ Include bins for separating dirty laundry by type (cold water wash, delicates, work clothes).

➤ Add a fold-out ironing board.

In smaller living areas, such as condominiums and townhouses, even more imagination is needed because of space constraints. One answer is a stacking washer and dryer, which take up less floor space than a conventional size appliance but can accommodate only smaller loads of wash.

Plumbing the Depths

One of the biggest space wasters in a laundry room is the typical laundry or utility sink. The old ones are huge, heavy, and way more than anyone needs. In our last house, we installed a basic stainless steel kitchen sink in a tall cabinet. This allowed for storage under the sink and was much more attractive than a standard utility sink. We tucked the washer and dryer under the basement stairs and enclosed them behind bifold doors. It was a great use of space and concealed the laundry at the same time.

A Short History Lesson

There may be great dignity in labor, but not in wasted labor. Washing machines and dryers, as well as modern fabrics, have made laundering unrecognizable by past standards. Washing, drying, and ironing clothes at the turn of the century was a multiday process involving boiling kettles of water, a clothes lines for drying, and a small collection of heavy clothes irons heating on a stove.

Go back far enough, at least in the United States, and clothes were washed in streams while being pounded with stones or rocks to drive out the dirt. Wash tubs came along next, with water heated in pots or kettles over a fire and crude homemade soaps. Scrub boards replaced rocks, but all the washing and rinsing was still done by hand. The first automatic clothes washer came out in 1937. Automatic washer sales, along with most home-building related products, really took off after World War II. Today's washers are far more efficient and offer an array of washing options.

Before You Buy ...

A washing machine is a major plumbing-related appliance. Unless you're single and living alone, it will be used by more than one person, who may treat it differently than you. Calculate your household needs when you go looking for a washing machine, and purchase one that will meet these needs:

➤ Be able to handle the amount of laundry you go through in a week

➤ Fit in your *laundry* area

➤ Have the washing options that you need

➤ Have an acceptable warranty

After you install your new washing machine, you'll have new considerations, such as its maintenance and safe, economic usage.

What's That Thingamajig?

The word **laundry** derives from *launder,* which started out as the Latin *lavare,* meaning "to wash." Along the way, it became the Old French *lavandiere. Lavatory* also derives from *lavare* by way of *lavatorium.* Those Romans were really into plumbing terms.

Safety Counts

It's easy to argue that we're becoming a society with excessive, even absurd, consumer warnings, but that doesn't mean that they're all illegitimate. Follow these rules for safe laundry operations:

➤ Don't wash or dry anything flammable, such as rags soaked in gasoline—the vapors can ignite or even explode.

➤ You know this one: Don't let your kids play on, around, or in your washer or dryer.

➤ Wait until the washer's agitator has completely stopped before reaching inside the tub.

Now that you know the safety guidelines, how about some for efficient usage?

Plumbing Perils

Soak any rags containing solvents or flammable liquids in a bucket of water, and allow them to air dry before washing them. Put the rinse water in a sealed container, and dispose of this safely. Better yet, consider soaking the rags and then throwing them away as well.

Laundry Hints

Efficient use of your washing machine and dryer doesn't have to crimp your style. It just requires a little planning and time management, buzzwords that have been floating around in business circles for years. Some good laundry practices include these:

➤ Wash in cold water with cold-water detergents when practical.

➤ Adjust the water level to the size of the load.

➤ Wash full loads as often as possible.

➤ Don't dry your clothes any longer than necessary.

➤ Dry heavier items (towels and jeans) separately from lighter, faster-drying clothes.

➤ Clean your dryer's lint filter after every load.

➤ Check and clean out your dryer's vent system periodically to ensure that it doesn't get blocked by lint.

While the clothes are getting cleaned, the clothes washer is getting a little scruffy inside and out. After each wash, wipe down the outside of the lid and the washer top. Wipe any soap residue from inside the tub with a damp rag and mild cleaner; do the same for the various dispensers (soap, bleach, fabric softener). Check the hoses and replace them if they look cracked or are developing bulges.

Pipe Dreams

Charge everyone, including yourself, $1 for the washer and dryer, and 50 cents for the detergent. No cheating—you have to pay each time. Pay $2 to wash a couple pairs of jeans, and you'll have a whole different attitude toward laundry usage. If your kids do their own wash with their own allowance, they'll start conserving in a hurry.

Everyone Has a Study

The Multi-Housing Laundry Association (MLA) is a professional group of laundry service providers and manufacturers that promotes common laundry rooms for multiresident property owners. They advocate one room full of happy coin-operated machines vs. individual washing machines and dryers in each apartment or living unit. I suppose that every industry has to have its promotional associations.

According to the MLA Water Usage Survey, a 90-day study involving more than 1,500 Phoenix, Arizona, apartments, laundry rooms are more than three times as efficient as individual laundry facilities in multiresident buildings. Why? When tenants had to pump all their spare change into the washing machines and dryers, they were much less willing to run small loads of wash. The survey estimated that each individual washing machine used an average of 11,767 gallons of water annually vs. an average of 3,270 gallons per coin-operated machine per apartment unit served.

Besides using additional water, the extra loads of wash consume gas and electricity to run the machines and generate higher sewage rates of waste water. From a building owner's perspective, a laundry room is certainly cheaper than installing individual machines in each and every apartment. The pertinence of this survey to a home-owner, however, is simply to show how efficient use of your laundry appliances can save you money and put less demand on your local utilities for power, water, and sewage disposal.

Pipe Dreams

If you want even more efficient use of your washer and dryer, find out from your utility companies if they offer cheaper rates at any specific times during the day, and do your laundry then, if possible.

Plumbing Perils

Run your dishwasher or clothes washer while you're at home rather than while you're out. If a hose breaks, you'd obviously want to be around to turn off the water. It happened to me once as I was literally walking out the door: A dishwasher over-flowed and began dripping into the basement. Had I left 60 seconds earlier, it would have gone on all day.

Location, Location, Location

Laundries often were located in basements by design and for practicality: If they leaked, damage would be at a minimum, especially if a house had a basement floor drain. First- and second-floor laundry locations are a great convenience, but they pose more of a danger if a hose ever bursts unnoticed. You can take a couple preventative measures in the event of flooding, however.

Water has to go somewhere when a pipe or hose leaks, and the preferred somewhere is into a drain rather than across your floor and down the hallway. Some older homes, whose basements were built in a truly utilitarian manner, have floor drains for errant water from washing machines or problem hot water tanks. If you're installing a laundry on a first or second floor, a floor drain will catch any water from either the washing machine or its hot and cold water supply hoses, if they leak.

An alternative to a floor drain is the FLOODSAVER, from AMI Inc. (1-800-929-9269 or www.floodsaver.com). This is a two-piece plastic washing machine surround with a drain pan on the bottom and a second section that attaches to or is supported by the wall behind the clothes washer. The hose bibs attach through the plastic, and the clothes washer sits in the pan. Any leaking water will go to the pan and into a standard shower drain. The manufacturer states that plumbing inspectors often allow the pan's drain line to run outside the house because it's only for emergency use. The FLOODSAVER sells for around $125, a bargain compared with the price of cleanup and repair after a flood. The FLOODSAVER fits most but not all washing machines.

Laundry Installation Rules

Washing machines and utility sinks have a few special plumbing requirements. These are mostly related to drainage due to the speed and volume of water that a clothes washer expels. A washing machine comes with three hoses attached to it:

➤ A cold water supply hose

➤ A hot water supply hose

➤ A discharge tube

The easiest setup is to have the washer and the utility sink close to each other because they both need hot and cold water supply pipes. The clothes washer, along with the dishwasher, should have air hammer arresters (some plumbing codes require them). The thinking behind this requirement is that an appliance using an electric solenoid device dispenses water much faster than a normal valve, causing water hammer in the pipes. Water hammer, as you may recall, occurs when water flowing through a pipe suddenly is stopped when a valve is shut off. The energy of the flowing water has to go somewhere, and it bangs against the pipes, which begin to vibrate, shaking the walls along with them. Severe water hammer can damage pipes.

To prevent water hammer, either air columns or air hammer arresters need to be installed. The simplest solution is to use a screw-on arrester that's installed between the faucet and the washing machine's hot and cold water supply hoses. The more complicated way is to tear open your wall and install an air chamber or a mechanical arrester on the affected pipes.

Pipe Dreams

If you do have a basement floor drain, test it before you need it by flooding it with water from a garden hose. Have the drain snaked out if the water drains slowly or backs up. You want to know that the drain is working when an emergency comes up.

Plumbing Perils

Check your hot and cold water hose bibs to be sure that they completely shut off the water to your washer. In case a hose bursts, you want to be able to shut off the water quickly. If either one is deteriorated and does not shut off completely, replace it.

Water In, Water Out

Getting water into a clothes washer doesn't require any elaborate plumbing: Install a tee in each of the hot and cold water supply lines and then whatever additional pipe is necessary to supply your two hose bibs. The bibs must be close to the machine for the hot and cold water supply hoses to connect to them.

Hose bib.

Mansfield Plumbing Products, Inc.

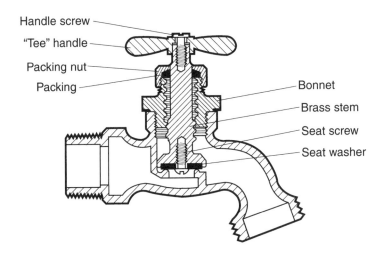

Handle screw

"Tee" handle

Packing nut

Packing

Bonnet

Brass stem

Seat screw

Seat washer

Getting water out is another issue. The discharge tube is a thick rubber hose that can empty into any 2-inch-diameter drain. You have several options for drainage, including these:

➤ A standpipe

➤ The utility sink

➤ A floor drain

A standpipe is an open-ended 2-inch-diameter pipe in the wall connected to the drain pipe. Think of it as a sink drain, but without having the usual stopper or strainer. The standpipe is left open at the top to accommodate the discharge hose. The height of the standpipe should be about level with the washing machine lid.

Other Drains

A utility sink will have a 1½-inch diameter drain and P-trap arm, but will drain into a 2-inch pipe in the wall downward from this drain branch. A 2-inch drain pipe has the capacity to handle eight fixtures. In this case, the clothes washer and the utility sink count as two units each, for a total of four. It isn't uncommon to see a discharge hose slung over the edge of a utility sink with the end near the sink's drain. Sometimes it's secured to the faucet with a small piece of bailing wire so that it doesn't slip out of the sink.

This is a perfectly acceptable means of draining a washing machine. The sink has enough capacity to hold the wastewater until it drains out the drain hole. It is not as neat of a setup as using a standpipe, however, and might be a result of budget considerations during construction or remodeling.

A floor drain also can serve as a washing machine's drain, but it's a little tacky.

Plumbing the Depths

It isn't your imagination that your socks disappear in the wash. Appliance repair technicians find clothes and spare change in washing machines and clothes dryers all the time when they disassemble them for servicing. Occasionally, a small piece of clothing can get caught in the discharge tube. If the machine is draining slowly, check the tube. You just might find a sock!

Electrical Requirements

A washing machine requires a dedicated 120-volt, 20-amp circuit. Adding a circuit to a service panel requires a permit from and inspection by your local building department. You shouldn't try to get by using an existing electrical receptacle that is part of another circuit, especially one that isn't grounded. Upgrading or installing a dedicated circuit for your clothes washer is just another example of a plumbing job overlapping into and requiring other work to do the job properly.

A Word About Clothes Dryers

The only time a dryer requires plumbing work is when gas piping is involved. This is a job for a professional. For that matter, your local code could prohibit you from doing any work with gas piping.

Dryers also require a 240-volt, 30-amp, dedicated circuit. Either you have an existing dryer circuit or you don't (an existing circuit will always be dedicated). Dryers plug into special appliance receptacles and cannot fit in standard receptacles.

Hot, humid exhaust heat from the dryer is discharged out a 4-inch-diameter duct that is usually made out of flexible plastic. Sometimes this duct connects directly to the outdoors by a small plastic vent, and other times it connects to a plastic or aluminum duct running between the floor joist. This duct then connects to the vent.

Plumbing Perils

Dryer lint can be a fire hazard if it's allowed to accumulate. Be sure to clean out the removable screen after every load is dried. Check the vent at the end of the duct as well. If the whole duct looks clogged, detach it from the dryer and blow it out with a garden blower.

It's imperative that your dryer vent to the outdoors! You don't want all that excess moisture coming into your basement or crawlspace. At best, it will promote the growth of mold and mildew; at worst, it will damage drywall and structural framing.

The Least You Need to Know

➤ The most important plumbing requirement associated with clothes washers has to do with the drain lines.

➤ Scheduled use of your laundry, using full loads, can be a real money and energy saver.

➤ Drainage precautions should be taken with first- and second-floor laundry rooms.

➤ Check whether your code requires water hammer arresters for your washing machine pipes.

➤ Laundry rooms must follow electrical code as well as plumbing code.

Chapter 25

Heat and Plumbing Concerns

In This Chapter

➤ Gas piping

➤ Fun gas appliances

➤ Steam and hot water heat

➤ Simple maintenance

Many new homes are heated with forced air gas furnaces, and that means gas piping. Gas pipe is typically black iron, and its fittings require their own version of Teflon tape (the yellow variety). With direct-venting, you can install gas appliances just about anywhere you can run a gas pipe. Natural gas and liquid propane are popular sources of energy that have been readily embraced by consumers.

A gas furnace operates on the principal of pulling in cold air, heating it, and distributing it throughout your home by blowing it through a series of ducts. Steam heat and hot water heat only secondarily heat the air, but they primarily warm objects through radiant heat from radiators. These heating systems require oil or gas fired boilers. They are entirely different systems, and each has its supporters.

There isn't much of your own work that you can do when it comes to gas pipe installation or an in-depth evaluation of your steam heating system. The former often runs up against code restrictions, and the latter is a mechanical art form. You can do rudimentary maintenance and pencil in a pathway for gas pipes on your remodeling plans. This chapter introduces you to steam and hot water heating systems, as well as gas piping.

It's a Gas!

Installing piping for a gas line is similar to the procedure for water supply pipes, but with a few exceptions. For one, you're only going to install a single supply line instead of hot and cold supply pipes for a fixture. There are no DWV pipes to install. And, oh, yes, gas can explode, while water can just drip or flood. Given the choice, most of us would choose the flood.

Gas pipe must be sized just as water pipe is sized. The number of appliances supplied by your gas line and their distance from the meter will determine the pipe size. Each appliance will have a different *Btu* per hour demand, which also must be taken into consideration. It's good to know this for basic information, but you'll probably never, ever install your own gas pipe, even if your local code allowed you to (it's unlikely that many will).

What's That Thingamajig?

Btu stands for British thermal unit, which measures heat. This is the amount of heat needed to raise the temperature of 1 pound of water by 1° Fahrenheit.

Your local municipality will issue a permit and do followup inspections and pressure tests for gas piping. The permit process takes varying amounts of time, depending on the responsiveness of your building department. Your gas company will extend a service line from its main pipeline on the street. This line will end at your gas meter.

An alternative to natural gas, if it isn't available in your area, is propane gas. Propane is stored in either an above-ground or an underground tank.

Plumbers install gas pipe, but they do not install the various appliances that use gas. Even if you cannot install your own gas lines, you can plan for and accommodate the installation of all kinds of fun gas appliances such as these:

➤ Furnace

➤ Range and cooktop

➤ Barbecue

➤ Fireplace

➤ Water heater

Plumbing Perils

If you have a preference for the location of your gas meter, discuss it with your gas company before the installation begins. You want it in a location that suits you, not the convenience of the installer. Building and plumbing codes may influence its location as well.

Many cooks swear by gas burners for their kitchens because they offer more control of the heat. Turn the knob, and you get the temperature you want immediately, no waiting for an electric burner to heat up or cool down. A natural gas barbecue means never, ever running out of fuel. Gas fireplaces are commonly installed in new homes, as are gas water heaters. You may be a mostly electric home now, but you don't have to stay that way if your house is near a gas line.

Gas Furnace

Like other modern furnaces, gas models have shrunk considerably in size from the old basement-domineering octopus-style furnaces. New furnaces are the size of a refrigerator and are much more efficient than past models. If gas is available in your neighborhood, you can replace your present oil or electric furnace or other appliances with natural gas models.

When choosing a gas furnace, keep these tips in mind:

➤ Have a qualified heating and cooling contractor determine the correct size unit for your house. (Tip: Bigger isn't better.)

➤ You can probably vent the furnace through your old venting system, but not so with many high-efficiency furnaces.

➤ Look for a model with a high-efficiency rating as indicated by its Annual Fuel Utilization Efficiency (AFUE); the minimum federally mandated efficiency level is 78 percent.

Aside from plumbing considerations, installing a new gas furnace (or any new furnace) might require a change in duct work as well as the cost of removing and hauling off your old unit. Ducts that are covered with asbestos will incur additional removal costs due to the asbestos abatement.

Pipe Dreams

Regardless of the type of furnace you have, a regular preseason inspection will help ensure smooth running when the weather turns cold. A technician will check the thermostat, air filter, blower, burners, heat exchangers, controls, and venting.

Barbecue Time

A natural gas barbecue requires that you run a gas branch pipe to the outdoor location of your barbecue. The barbecue itself connects up to the line with a quick-release connection. A shutoff valve is installed near the end of the branch line so that the gas can be conveniently turned off when you leave town for an extended vacation, store the barbecue away for the winter, or need to remove the barbecue for any other reason.

Some of the advantages of a barbecue with piped natural gas include these:

➤ No more propane tanks

➤ Lower cost than propane gas

➤ An available cook space if your electricity goes out and you have an all-electric kitchen

A quality natural gas barbecue doesn't come cheap and can run as much or more than a kitchen range. Then again, you'll never buy charcoal again.

Fireplaces

There's nothing like a wood burning fireplace—once you get the fire going. Let's face it, starting a fire is a nuisance no matter how much of a back-woods-person you might be. Why bother when you can flick a switch and let natural gas do the job?

Pipe Dreams

Inquire with your gas company if you're even considering converting to gas or adding gas appliances. Some offer rebates when you install new gas appliances or simply adapt to gas.

Gas fireplaces come in two flavors:

➤ Those that use only gas and man-made logs

➤ Those with a gas-fed grate for wood burning

A fireplace with fake logs is the easiest arrangement, but for you purists, you can cheat with a gas grate that burns regular logs. I grew up with these fireplaces, and I can testify that any log, regardless of how unseasoned the wood or snow covered it might have been, would eventually burn on a gas grate. Existing-wood burning fireplaces can be converted to gas with the installation of a gas grate after a gas pipe is installed.

New gas fireplaces with man-made logs vent directly through an outside wall and do not require a chimney. Protective, heat-proof glass in place of a fireplace screen ensures your safety and keeps room heat from going out the flue.

Plumbing the Depths

Natural gas is colorless and odorless and made up mostly of methane. A chemical odorant is added so that we can smell it and detect leaks. Natural gas is a clean-burning fossil fuel that releases carbon dioxide and water vapor during combustion. When the gas burns incompletely with an insufficient air supply, it produces carbon monoxide, another colorless, odorless gas, except that this one's toxic. The United States is the world's biggest consumer of natural gas, and we produce most of the gas that we use. The first successful domestic gas well, which was all of 27 feet deep, was drilled in Fredonia, New York, in 1821.

Gas in the Kitchen

The big advantage of natural gas ranges and cooktops is temperature control. Gas offers instant-on and instant-off heat. Automatic ignition systems eliminate pilot lights.

If you're not ready to install a gas range but are having other gas pipe work done, consider running a pipe through your basement or crawlspace and stubbing it under your kitchen floor for future use. Just be sure that you know the future location of the appliance.

Water Heaters

If you want to learn about gas water heaters, refer back to Chapter 23, "Hot Water Heaters."

Gas Space Heaters

Another option for gas heating is the installation of individual gas space heaters. These can provide heat to individual rooms or new additions when running a furnace duct is impractical but running a gas pipe is not. Of course, you should do some cost comparisons between installing a gas space heater and installing an electric one. You or your heating contractor will have to calculate your heating needs and determine which system can best do the job.

Individual gas heaters and fireplaces come in two models:

➤ Vented

➤ Unvented

Vented models need access to the outside of your house, of course, meaning a chimney or a direct vent out the wall. An unvented model can be located anywhere in a room because it doesn't have the limitations a vent presents. This convenience has some drawbacks that are pointed out in the next section.

Pipe Dreams

Instead of installing an individual gas space heater, consider a gas fireplace with a built-in blower. A fireplace is more handsome than a space heater and is a good selling point when you sell your home. Blowers are very effective, and a single fireplace can heat a large area.

Is Ventless Also Witless?

A ventless gas heater (or fireplace) has no direct vent to the outside. They are legal in most states, but not in Canada. As with most building practices, your local code makes the final determination, so always check with your building department.

A vented appliance is attached to a duct, chimney, or pipe to carry away these byproducts of the natural gas combustion:

➤ Carbon monoxide

➤ Nitrogen dioxide

➤ Sulfur dioxide

An unvented heater is designed to burn at close to 100 percent efficiency. It releases any remaining trace byproducts into your house. Built-in oxygen depletion sensors determine whether the available oxygen in the room is decreasing and automatically shut down the appliance. Various natural gas industry representatives tout studies that show unvented appliances to be safe and that they do not create excessive amounts of indoor air pollution.

Skeptics say, not so fast. If they're so safe, why aren't carbon monoxide detectors included with them? And what about the high levels of water vapor they produce? Who wants that inside their home?

Think this through for a moment: Is there any logical reason to install an unvented heater or fireplace in your home? Why expose yourself to possible contamination from burning gas, not to mention the water vapor, when a simple vent can eliminate most of those possibilities? I say *most* because a damaged vent can be a source of carbon monoxide poisoning, but how likely is this to happen?

Run your gas pipe for your new heater or fireplace, but skip the unvented models.

Vented Heating

These models need to be mounted on an outside wall to install their vents. Some fireplace models can use a chimney. These heating appliances come in three models:

➤ Baseboard heaters

➤ Space or room heaters

➤ Wall furnaces

Each has a maximum area that it can heat efficiently. Given the cost of running a gas pipe, it hardly seems worth it to install a single gas baseboard heater when an electric baseboard can be installed instead. A wall furnace that can heat a large area might prove to be an economical heating approach.

Hot Water and Steam Heat

At one time, most commercial office buildings and apartment houses were heated with either a hot water system or a steam system. It was just more feasible to circulate water or steam floor by floor and maintain a comfortable winter temperature than doing so with forced air. For all of you who may have complained to building superintendents about stone cold radiators in the middle of December, this would more often be a matter of poor system maintenance or an undersized boiler rather than the principles of the system being faulty. Of course, this doesn't offer much comfort or solace when you see your breath vaporizing and you're sleeping in three layers of woolen clothing.

Hot water and steam heat were particularly suited for large homes, too. By the 1940s, forced air systems had surpassed hot water and steam in residential applications. They have retained their popularity due to cost, convenience, and ease of maintenance. Nevertheless, if you have an existing hot water or steam system, it's often well worth keeping in good operating order for its efficiency and comfortable heat.

Plumbing Perils

If you're buying an older home with steam or hot water heat, take a look at the maintenance records for the system. You want to be sure that it's had regular servicing. Look around each radiator for water stains under the valves. These can be indications of regular or at least recurring leaks. You'll want to know ahead of time whether you're looking at major repairs.

Finding Yourself in Hot Water

Hot water or hydronic heat starts with a gas or oil-fired boiler that heats water to around 180° Fahrenheit. Pumps move the water through pipes to radiators. Hot water radiators are smaller than steam radiators, and the system itself is comparatively quiet. Hot water heat is now used in place of steam in homes and smaller buildings.

Some of the advantages of hot water heat include these:

➤ Ability to individually control the temperature of each room that has a radiator (just adjust the valve)

➤ Long life of components

➤ Efficiency

➤ Freedom to locate the boiler away from heated areas

➤ Dust control

➤ Less room needed for the pipes than hot air ducts

Modern hot water heating uses baseboard-style heating units instead of the old-fashioned cast-iron radiators. Both require pipe to carry hot water to the heating unit as well as back to the boiler.

The Boiler

The boiler used in hot water heating systems isn't much different in principle than a tea kettle heating up on your stove. A boiler circulates hot water through a header pipe and a flow valve to your radiators or baseboards. A circulator pump then returns the water to the boiler to be heated again. A pressure-reducing valve reduces the water pressure that comes into the tank from your water supply pipe. An aquastat regulates the boiler's temperature. A pressure relief valve prevents the buildup of excess pressure.

New high-efficiency boilers are available that reach an efficiency of close to 85 percent, compared to 60 percent or less for older models. These boilers are vented directly out a nearby wall.

Understanding Steam Heat

Steam systems predate hot water heat. Their boilers reach high enough temperatures to convert water to steam that then rises through pipes to radiators. Steam heat is noisier than hot water heat and doesn't always heat a building evenly.

Plumbing Perils

Old steam pipes are typically wrapped with asbestos insulation. Any pipe without insulation can allow steam to condense before it reaches some radiators, thus rendering it ineffective. If you have steam pipes and a previous owner removed the insulation, be sure to install replacement insulation (this will not have any asbestos).

Steam pipes have to be installed for the most efficient operation of the system. This is especially true for single-pipe systems in which the steam and the condensation must share a pipe. Steam mains and the horizontal pipes are installed at a certain pitch to allow the water to pass by the steam without affecting it. If the pitch of the pipes changes due to building settling, the condensation can collect in various portions of the pipe, cool, and condense any steam passing through it. This results in uneven heating.

Steam moves very easily through a system of pipes to your radiators. It also has a high heat-holding capacity (higher than hot water). Steam moves through the system's pipes and then condenses after it heats the radiators and returns to the boiler as water.

Radiators heat people and objects in a house by radiating heat outward. They also heat the air that then moves by convection currents.

One Pipe or Two?

The simplest steam heat is a one-pipe system. A single supply pipe brings steam to the radiator and acts as a return pipe for the condensation as it returns to the boiler. Single-pipe systems use large pipes, often greater than 2 inches in diameter, to accommodate both the steam and the returning water.

In a two-pipe system, a second pipe carries the condensation back to the boiler.

Plumbing the Depths

Early American heating consisted of fireplaces and wood-burning stoves until the advent of hot air furnaces in the early 1800s. Furnaces advanced the use of steam-heating apparatuses, and the city of Philadelphia became the center of heating technology into the 1830s. High-pressure steam boilers were subject to explosions unless they were constantly monitored, so the main marketplaces for these heating systems were large buildings. Improved technology in the mid-nineteenth century allowed smaller buildings to go with steam heat. Eventually, entire steam lines were laid down under city streets to supply steam heat to industrial, commercial, and residential buildings.

Bleeding Your System

Two types of bleeding are involved with hot water systems. One is a total system bleed, and the other is done radiator by radiator. You bleed them so that they'll keep working efficiently.

When air gets trapped in hot water radiators, it displaces the water. Without the water, that section of the radiator won't heat up properly. Each radiator should be bled once or twice a season. After placing a pan under the radiator valve to catch some of the inevitable water that will flow out, you use a radiator key to open the valve and bleed the air out. You keep the valve open until only water comes out.

A hot water or steam-heating system should be checked every year or two by a qualified technician. Look at it as cheap insurance. Have your technician show you how to bleed the system to get rid of any air pockets in the pipes and contaminated water. Bleeding the entire system is normally a simple procedure (do it when the system is cold and not being used):

Pipe Dreams

Schedule your hot water or steam heat system inspection prior to the start of the heating season, such as during the late summer. It's a lot easier to get a technician to come out when you don't need the heat and you're not competing with customers who have lost theirs.

351

➤ Close the shut off valve above the circulator pump.

➤ Attach a garden hose to the drain valve, and run the other end into a drain or outside.

➤ Open the drain valve.

The designs of each manufacturer will vary, and older systems might be a little trickier to bleed, so have your technician run through it with you. It could be that some of your valves are in need of replacement, and you'll want them attended to before you try to open them.

Plumbing the Depths

Another form of hot water heating is radiant flooring, which turns your floor into one big radiator. Flexible pipe is either laid down and covered with the concrete slab or is snaked through joist and installed under a wood floor. Radiant heating pipe must be properly sized according to the house dimensions, insulation in the walls and roof, and the amount of window exposure. Carpet and specific types of floor coverings will also affect the heating results. Radiant flooring got a bad reputation when at least one major installer's systems leaked, with the inevitable lawsuits that followed.

Radiant Heat Through Radiators

Old cast-iron radiators weigh a ton but are great distributors of heat. You can even get replacement radiators, although new models are slimmer than the originals. New flat-panel radiators are also available. Cast iron radiators offer both radiant heat and convection heat. Specially made disclosures can hide your radiators if you don't like the looks of them. Metal enclosures are best; wood enclosures act as an insulator, retaining heat as well as possibly warping or splitting from the effects of the heat. Radiators with multiple coats of paint on them can be less efficient at heat distribution because the paint can act as an insulator.

Most old radiators were freestanding in front of windows, but some were recessed and built into the walls for a nice finished touch.

The Least You Need to Know

➤ It's likely that your local code permits only licensed plumbers to install gas pipe.

➤ Running a gas line through your house isn't much different than running a water pipe; your plumber will go through the same motions of measuring, drilling through wood, and installing as would be done with water pipe.

➤ Old steam and hot water heating systems need regular servicing and inspections to keep them in working order.

➤ Think twice before installing an unvented gas heating appliance.

353

Chapter 26

Wells Are Swell

In This Chapter

➤ Our dependence on ground water

➤ The basic system

➤ You're responsible for your water

➤ Disinfecting your well

➤ Safety precautions and maintenance

If you're an urban dweller, about the only connection you have with a water well is a fake wishing well at an equally fake Bavarian theme restaurant. In real life, a well owner would be very annoyed if you threw your spare change into the family water supply. Some statistics show that just over half of the American population depends on some form of ground water—both from private wells and municipal water supplies—for its potable water supply.

In the old days, at their most basic, a well was built the hard way: Grab a pick and shovel, and start digging. This got to the water, but it wasn't the safest construction model. A "dug well" is more prone to contamination and deterioration than more modern drilled wells. Aside from that, it's a lot easier to use a drilling apparatus than a pick and a shovel.

This chapter covers some of the basics on well construction, maintenance, and sanitation. When you own your own well, you control your water supply. You never pay a water utility for your water, but you don't get their system maintenance, either. You'll need to know how to keep it clean to keep it safe for you and your family.

Who Drinks Well Water Anyway?

For those of us who live in cities, the idea of a well in the backyard is about as likely as having a pigsty or a chicken coop in that same yard. Our water comes from a city water system and food from the grocery store. This may not make us the most independent folks around, but it beats cleaning up after a brood of chickens. Nevertheless, ground water supplies a significant amount of potable water to both individual homes through the use of wells and community water systems that tap this same water.

Plumbing the Depths

Wells have played a part in history and mythology as well as simply supplied drinking water. The holy wells of Ireland have a significance going back to prehistoric times. The Celts, an earth-centered culture, believed that certain places, such as the holy wells, possessed curative powers. After St. Patrick and Christianity took root, the ancient rituals surrounding the wells were replaced with Christian symbols and rites. No one seemed too concerned about bacteria levels in the water, however.

Ground water from deep wells is filtered naturally through layers of rock and sand and is far less likely to become contaminated than surface water. Worldwide, after you exclude the polar ice caps, ground water makes up 90 percent of all fresh water, according to the National Groundwater Association (NGWA). Aquifers or underground reservoirs and artesian wells are some of the largest sources of this fresh water.

What's That Thingamajig?

Ground water is the result of rain and snow that soak into soil and move downward, accumulating inside any openings inside beds of rock and sand.

The National Well Owners Association (NWOA) states that ground water supplies the United States with close to 20 percent of its daily water needs, including agriculture and industrial. California alone pumps more than 14 billion gallons of ground water a day. An estimated 8,000 water-contracting firms drill close to 800,000 bore holes each year in the United States, according to the NWOA. Ground water advocates suggest that ground water is greatly underutilized despite our nearly 16 million wells. By contrast, China has close to three and a half million wells serving a much larger population. Our largest use of *ground water* is for irrigation.

A private well puts the responsibility for water quality in your hands. Surveys indicate that well owners prefer the taste and low cost of ground water over that of municipally supplied water.

Ground Water Groups

There's a club, association, or society for just about every interest, and ground water users and advocates are among them. Their groups include these:

The National Ground Water Association
601 Dempsey Road
Westerville, OH 43081
1-800-551-7379
www.ngwa.org

National Well Owners Association
www.wellowner.org

EPA Safe Drinking Water Hotline
1-800-426-4791

The American Ground Water Trust
603-228-5444

Water Quality Association (WQA)
630-505-0160

Pipe Dreams

A deep well is far less affected by droughts than surface water supplies. Drill your well as far down as is practical, given the soil conditions and water table, to ensure a constant water supply.

Each of these organizations can assist you with your ground water questions and concerns.

The First Time

You might be contemplating drilling a well on your property or purchasing property that will need a well. It is paramount that you find your water supply before you start building, for two reasons:

1. You need to know that you have an adequate supply of water.

2. You don't want to discover that your only source of water will require a well to be dug where you had planned your new kitchen.

Drilling a well is specialized work and may require a permit (always check first). Contract with an experienced, licensed firm to do the drilling. Keep in mind that you might hit dry holes or a low-yield area, so more than one hole can be required. If you are considering the purchase of raw land, write a contingency that the purchase is dependent on your finding an adequate water source before the purchase and sales agreement is finalized. It's a lot cheaper to drill a few test holes first than get stuck with worthless land. Your prospective neighbors can be very helpful in this area by sharing there drilling experiences with you. Any local contractor will probably be aware of water conditions as well.

Your well should be situated where it's easily accessible for cleaning and maintenance, but at certain minimum distances:

➤ Fifty feet from a septic tank or tile sewer

➤ Ten feet from an iron sewer line

➤ Five feet from a property line

Your local codes might require different distances, so check with your building department before you begin constructing a well.

More Than Just a Hole

A private water system that's supplied by ground water must be planned and constructed around the following:

➤ An adequate, dependable supply of clean water

➤ The depth of the well

➤ A pump and a pressure tank

➤ A power source to run the pump

➤ The size and length of the pipe between your house and the pump

Plumbing the Depths

In truly rural areas or on sections of property that don't have nearby electrical power, internal combustion engines or windmills supply power to run pumps. These systems work well for limited agricultural purposes, but not for supplying a home with fresh water.

It's beyond the scope of this book to detail the complete installation of a private water system (there are entire books written on it, I'm sure). The basic system works like this:

A well is dug, and a source of ground water is secured.

A pump is installed as close to the well as possible (submersible pumps are installed in the well itself).

A power source is connected to the pump.

Pipe is run from the well to your house.

A pressure tank is installed.

The pump runs on and off, filling the tank, as you draw water from the faucets.

Pump and tank sizes are coordinated with the capacity of the well to provide water. Your local contractor can help you determine the system that's best for you.

Plumbing the Depths

A 1994 water quality survey covering nine Midwest states and carried out by the Centers for Disease Control and Prevention produced some less-than-cheery results. Coliform bacteria was found in more than 40 percent of the 5,520 wells tested, and E.coli was found in 11 percent of the wells. Older dug wells showed the greatest contamination while newer, deeper drilled and driven wells showed less. Cracked casings also increased the occurrence of contamination. This survey makes a strong case for regular testing of well water.

Types of Wells

You are probably familiar with a dug well. These are shallow wells with built-up walls and dug at low points on a property. They are subject to surface water contamination as well as subsurface seepage. Insects also are a problem due to the comparatively large diameters of the wells and the amount of water consequently exposed.

A *drilled* or bored well, on the other hand, is much narrower and has a casing or pipe lining its upper section. These wells are deeper than dug wells and can be contaminated if casings crack or rust, and they also allow subsurface water to seep in. Casings that do not extend far enough above the ground can allow surface water to get into the well and cause contamination.

What's That Thingamajig?

A **drilled well** consists of a deeply bored hole with a casing lining the upper section. The casing is a strong pipe that prevents the sides of the hole from caving in. Water enters the well from below the casing.

As another preventative measure, a well should have a cap or seal to cover the opening and prevent debris, insects, and other contaminants from finding their way into the water. If ground water levels rise above their normal level (after a lengthy period of rain, for instance), they can reach upper soil levels, where bacteria can then enter the water.

Contracts and Contractors

Hire a drilling contractor or a well maintenance firm the same way you would hire a plumber:

➤ Get recommendations from friends and neighbors.

➤ Check for a license, bond, and insurance.

➤ Inquire about certification, such as the Master Ground Water Contractor (MGWC) from the National Ground Water Association.

➤ Look at the contractor's past work.

Always use a written contract spelling out what work will be performed and that all work will comply with local and state regulations. The contract should state the size and thickness of the casing and the types of screen and well cap to be installed, as well as the disinfection procedure. Any and all warranties on both work and materials should also be included in the contract.

After the well has been completed, check the depth, water yield, and cap or seal (it should be at least 6 inches above the level of the surrounding ground). You also should receive a copy of the construction record; file this away with your other house papers.

Plumbing Perils

Abandoned wells are a problem from several perspectives. They act as direct routes for surface pollutants to reach ground water and are a safety hazard for people and animals who can tumble into them, making the property owner responsible. All abandoned wells should be properly sealed.

Abandoned and Forgotten

Unlike the myriad consumer goods we all go through and eventually toss away once they're outdated, a property owner can't exactly turn a blind eye to an abandoned water well. Even if you ignore the danger to yourself or the well as a conduit for pollutants, there's always someone named Timmy falling down one of these wells while his dog Lassie runs around figuring out what to do. An abandoned well is one that no longer supplies potable water or that is too deteriorated to allow access to ground water. Some estimates of the number of abandoned wells run into the hundreds of thousands.

You should seal your abandoned wells for the following reasons:

➤ To protect the ground water. The well can become a conduit for surface pollutants to reach the ground water.

➤ To prevent humans and animals from falling into the hole.

➤ To comply with local laws that probably require it.

A sealed abandoned well provides assurances that let you sleep better at night.

Well Maintenance and Safeguards

Along with the independence of your own well comes the dependence of doing or managing your own repairs and upkeep. Some of these chores include the following:

➤ Conduct an annual test for bacteria and other contaminants in the water. Test your well any time the water tastes, smells, or looks different than normal.

➤ Keep all hazardous house, yard, and automotive chemicals (pesticides, paints, oil) far away from your well.

➤ Be sure that any new construction on your land, waste systems, and chemical storage facilities are constructed the proper distances from your well.

➤ Regularly check the well cover to be sure that it's in good condition.

➤ Keep the top of your well at least 6 inches above the surrounding ground (1 foot is better).

➤ Hire only licensed contractors to work on your well and pump.

➤ Maintain records of all repairs and testing.

Yearly testing is a big deal and should not be neglected. Your health depends on it!

Testing, Testing

A water quality test should be performed yearly or as needed. This latter term means when the water suddenly smells, tastes, or looks bad. Testing will check for the following:

➤ Contaminants, such as bacteria, ammonia, and arsenic

➤ Hardness

➤ Iron bacteria

➤ Other biofouling organisms

> **Pipe Dreams**
>
> Are you new to well water and having problems with your well? Talk with your neighbors. They've seen it all before, especially if they've been in the area for a while. They probably can advise you on solutions and good local contractors—as well as who to avoid.

Iron bacteria isn't harmful in and of itself, but it can cause staining of plumbing fixtures and laundry, as well as increased corrosion of pipes and pumps.

Be sure to use only state-certified testing laboratories (check with your local health department for referrals). If your tests show that your water needs to be treated, contact a qualified water treatment specialist. The Water Quality Association (WQA) promotes a voluntary certification program for water treatment specialists and can be reached at www.wqa.org.

Plumbing Perils

Don't be so sure that your well is the source of bacteria in your drinking water. A damaged plumbing system or one that is installed incorrectly can also be a source of bacteria. Discrepant results between water tested from your tap and water directly from the well might hint at a plumbing problem.

Pipe Dreams

One approach to ensure that your well water stays clean is to disinfect your well plumbing every time you service it or service the pump. Wash the system down with a chlorine solution after all your work has been completed.

Disinfection

For our purposes, disinfection is the act of disinfecting well water by removing infectious agents. Disinfectant agents vary by type and concentration. The standard methods for disinfecting well water are similar to those mentioned in Chapter 1, "Open Tap, Get Water," regarding water and sewage treatment:

➤ Chlorine

➤ Filtration

➤ Ultraviolet radiation

All these approaches require systems that are installed and maintained by professional technicians. No water is ever rendered 100 percent pure, nor should it be (some water-borne bacteria is actually beneficial). The very act of drilling introduces some contaminants to your well, but careful cleaning and disinfection will keep any problems to a minimum.

How Shocking

If your well water is contaminated with unacceptable levels of bacteria, your response should be twofold:

1. Decrease the bacteria to a safe level.

2. Eliminate the source of the bacteria.

A working well doesn't suddenly become contaminated on its own. The source of contamination might be the well itself acting as a funnel for contaminants or a leaking septic system. Look at your well construction. The casing should extend above the surface of the ground, and the ground around the casing should slope away from the well.

One accepted way to clean out bacteria in contaminated well water is through shock chlorination. I'd like to thank the Extension Service of Mississippi State University, whose work I have fully quoted, for the following information.

How Shock Chlorination Works

Shock chlorination involves introducing a strong chlorine solution into the water source and plumbing system, and letting it disinfect the system for 12 to 24 hours. You can use regular household bleach that contains 5.25 percent sodium hypochlorite, which kills bacteria and certain viruses.

Shock chlorination is recommended after a new well is constructed and installed, any time a well is opened for repairs, or if flood water has entered a well. *Be sure to store enough fresh water to last 12 to 24 hours while the well and water system are being disinfected.*

Shock chlorination is recommended for all bacterial contamination. Flushing after treatment is required to reduce the chlorine concentration; flush until all chlorine odor is gone. Also, use an inline cartridge water filter to remove iron or sulfur bacteria that may dislodge from plumbing lines during chlorination. Be sure to change the filter regularly to keep it from becoming clogged.

How to Use Shock Chlorination

To use shock chlorination effectively, you must introduce the chlorine solution directly into the water supply. Depending on how your well is constructed, some pipe disconnections or mechanical adjustments may be needed. If you need help doing this, contact a licensed well driller or plumber in your area.

First, clean the well, spring house, or storage reservoir. Remove debris, and scrub or hose off any dirt or other deposits on interior surfaces. Pump the well to remove any suspended solids or foreign matter in the water. Scrub interior surfaces with a strong chlorine solution of a half-gallon of chlorine laundry bleach (nondetergent and unscented) in each 5 gallons of water.

To know how much chlorine solution to use to disinfect your well, you must know the number of gallons of water in your well. This is determined by the diameter of the well casing and the depth of the water in the well. Follow directions in the first table that follows to calculate the number of gallons of water in your well.

Add chlorine bleach into your well until you reach a concentration of 200 parts per million (ppm). (Note that you should disconnect any activated

Plumbing Perils

Scented laundry bleach is fine for your clothes, but not for your well water. Follow the guidelines and use only plain, unscented bleach.

carbon or charcoal filters before adding bleach or chlorine; chlorine is also toxic, so wear gloves and goggles, and keep it off your skin.) Use plain laundry bleach.

Plumbing Perils

Follow the recommended bleach/ water mix when disinfecting your well. A small amount of additional bleach might be warranted, but not much. This isn't a case where more would be better. You want to be able to get it out of the system after the disinfection. Excess bleach will hamper that removal.

Water containing chlorine bleach is not safe to drink. Follow shock chlorination procedures carefully, and be sure that there is no chlorine odor before drinking the water.

This concentration should kill all bacteria. You can use the table on the next page to help you calculate how much chlorine to use to reach this concentration.

You can use other chlorine sources instead of common household bleach. Be sure to handle them with extreme caution because they are much stronger. Wear protective gloves, clothing (aprons), eyewear, and shoes. Chlorine is toxic and corrosive, and it can burn your skin or irritate your eyes. Rinse off any exposure immediately, and if irritation persists, consult your doctor.

The following table will help you determine the amount of water contained in a water well of varying diameter, for each 1 foot of depth.

Well Casing Diameter (Inches)	Water Per 1-Foot Depth (Gallons)
2	0.163
4	0.65
5	1.02
6	1.47
8	2.61
10	4.08

To figure the number of gallons of water in your well, multiply the gallons per foot by your well water depth. For example, if your well casing is 5 inches in diameter, there are 1.02 gallons of water for every foot of water. If your well water depth is 100 feet, your well contains 102 gallons of water ($100 \times 1.02 = 102$ gallons). Contact a well driller if you need help finding your well water depth.

The following table will help you estimate the amount of 5.25 percent chlorine bleach (nondetergent, unscented) needed for different amounts of water to equal 200 ppm.

Bleach to Add in Your Well (Pints)	Amount of Water in Your Well (Gallons)
¾	25
1½	50
2¼	75
3	100
3¾	125
4½	150
5¼	175
6	200

Common Conversions:

1 cup = 8 ounces

1 pint = 16 ounces

1 quart = 32 ounces

1 gallon = 128 ounces

After determining the amount of water in your well, use the second table to find how much 5.25 percent chlorine bleach to use to disinfect your well. You also may use other chlorine sources, but using household bleach might be easiest.

The mixed chlorine solution must be poured directly into the well. The best way to add chlorine to a drilled well is to fill a tank or other container that holds more water than is stored within the well casing. Mix the chlorine solution with the water in the tank, and then let the tank contents flow into the well. Or, put the required chlorine tablets in a weighted porous sack, and lower and raise it within the entire water depth until the tablets have dissolved. Some wells may require different dispensing methods, depending on their construction.

After adding chlorine, attach a hose to the nearest faucet. Turn on the pump to recirculate the chlorinated water. Use the hose to wash down the well casing and drop pipe as the water is returned to the well through the hose. For the process to be effective, the returning water must have a strong chlorine odor. If it doesn't, add more chlorine to the well. If you use common household bleach, be sure that the bleach is nondetergent and unscented.

Drain the water system accessories, such as the water heater, and refill with chlorinated water.

Pipe Dreams

Technology is always changing, so keep in touch with your water system contractor regarding easier ways to disinfect your well water. New, improved solutions and formulas might come on the market at any time for disinfection, and you might want to consider one of them rather than using chlorine bleach.

Release the air from the pressure tank (except for tanks with a permanent air cushion) to fill the tank completely with chlorinated water.

Before disinfecting the water lines leading to and inside of your home, temporarily remove or bypass any carbon filter in the system. Then, one at a time, open each faucet and hydrant in your home. Let the water run until it has a strong chlorine odor at each faucet. Add more chlorine solution at the well if the chlorine odor is not apparent at any faucet.

When the chlorine adequately reaches all faucet points, let the chlorinated water stand in the well and household plumbing at least 12 hours—preferably 24 hours—to kill bacteria completely. Open outside hydrants and let them run until all chlorine odor is gone. Next, flush all lines inside your home until all chlorine odor is gone. Follow this order of elimination to reduce the chlorine effect on your septic tank. Strong chlorine solutions can kill the bacteria that make a septic system work. Do not allow more than 100 gallons of shock chlorinated water to flow through faucets and drains that lead to a septic tank. Do not use this water for a garden or lawn.

Strongly chlorinated water will not harm livestock, but they will refuse to drink it unless very thirsty. Fill livestock tanks before starting to shock chlorinate if the strong chlorine solution is to be left in the system for a long period.

Follow-Up Treatment

After you have thoroughly pumped the well to remove the chlorine, use the water for a week and then have another water sample tested. Two or even three consecutive tests give you more confidence that the problem has been corrected. No bacteria test is perfect, and the results of only one test can be misleading. If tests continue to show bacteria, you may need to retreat and retest your water. You also may need to have your well site evaluated to help determine the source of the disease-causing bacteria or iron and sulfur bacteria problems.

Continuous Chlorination

If several shock chlorinations do not remove the bacteria, your well may require continuous chlorination to eliminate the bacteria problem. This requires equipment to add chlorine to the water as it is drawn from the well. The chlorine must be thoroughly mixed with the water and have time to kill all bacteria. This equipment may be purchased from a water-treatment company, a local well driller or a swimming pool dealer.

Summary

Shock chlorination is a good way to treat coliform, iron, and sulfur bacteria in your private well. Always find and correct the source of any bacteria contamination in or around your well. For more information on water quality, contact your county extension office.

References

Small Water System Operation and Maintenance. California State University—Sacramento, School of Engineering, Applied Research and Design Center. Adopted by Environmental Protection Agency, 1987.

Manual of Individual and Non-Public Water Supply Systems. Environmental Protection Agency, Office of Water, May 1991.

The following publication was produced with United States Department of Agriculture/Extension Service funds from grant number 90-EHUA-1-0014.

Publication 1865, by James G. Thomas, Leader, Extension Agricultural Engineering; Jack W. Carroll, Coordinator, Energy Extension Center; and Terry S. Holder, Extension Water Quality Program Assistant. Extension Service of Mississippi State University, cooperating with the U.S. Department of Agriculture.

This document is public information and may be reproduced in part or in total. It should not be used to imply endorsement of any specific brand or product.

Giving Your Well the Treatment

Once again, another co-op extension service steps in to provide an abundance of information on water treatment systems. This time I have Glenda M. Herman and Gregory D. Jennings (publication number HE 419) of the North Carolina Cooperative Extension Service to thank for their timely article.

Plumbing the Depths

The U.S. Department of Agriculture sponsors the Cooperative State Research, Education, and Extension Service (CSREES) that, in their own words, is an international research and education network: "CSREES links the research and education programs of the U.S. Department of Agriculture and works with land–grant institutions in each state, each territory, and the District of Columbia. In cooperation with our partners and customers, CSREES provides the focus to advance a global system of research, extension and higher education in the food and agricultural sciences and related environmental and human sciences to benefit people, communities, and the Nation." What a wonderful and useful program.

The authors quite rightly state that you must identify the type and level of contaminants in your system before deciding on a treatment procedure. Normal bacteria and other microorganisms, for instance, can be treated by standard disinfection with chlorine. Taste and odor problems will improve with a carbon filter, while rotten egg odor (hydrogen sulfide gas) requires an oxidizing filter followed by a carbon filter. Do you have too much sediment? Install a fiber filter. Hard water (calcium and magnesium) calls for a water softener.

Dissolved iron (a problem pointed out by my editor in *her* well) should clear up with a water softener (for up to 5 milligrams per liter), an iron filter, or chlorination followed by sand and carbon filters.

A carbon filter removes most organic compounds that produce taste and odor issues. The more carbon in the unit, the greater its effectiveness. They get replaced as soon as a foul taste or odor returns to the water. The most expensive carbon filter is a point of entry (POE) filter that treats all the water coming into your house instead of using individual filters attached to faucet spouts.

A fiber filter contains spun cellulose or rayon and removes sediment from the water. The finer the fibers in the filter, the smaller the particles it will trap (and the more frequently it will have to be changed). These filters do not purify water; they only remove, to a point, the visible junk that can contaminate it.

Plumbing Perils

Be sure to purchase a water treatment system that's responsive to your specific water problems, not a one-size-fits-all solution. Have your water tested by a lab that is not associated with a water treatment supplier. Get at least three estimates for the work, and be sure that the sales representatives are knowledgeable and competent about different treatment options.

Chlorine is the most common means for disinfecting municipal and individual water supplies in the United States. It oxidizes the internal enzymes of bacteria (kind of an internal toasting). If you install equipment that dispenses chlorine, it should be easy to use, should be virtually automatic, and should require infrequent maintenance while treating all water coming into your house. As an alternative to chlorine, iodine can be used, but it's more expensive.

Adjusting the pH

pH is a measure of the acidity or alkalinity of a solution—in this case, your well water. You want your water to be as close to a pH of 7, or as neutral as possible (less than 7 is acidic, and greater than 7 is alkaline). Water that is too high or low in its pH is corrosive and can damage pipes and fittings. Tank-type neutralizing filters or chemical-feed pumps that inject a neutralizing solution into the well are used to adjust the pH. Your local well contractor can show you each system and discuss its effectiveness relative to your particular water problems.

Water Softeners

Water softeners remove calcium and magnesium ions from hard water by exchanging them for sodium ions as water passes through an ion-exchange resin. Softer water is easier on your pipes and water heaters and any activity involving soap (the soap will dissolve more easily). If there is oxidized iron or iron bacteria in the water, however, the resin becomes clogged and can no longer soften effectively, in which case you'll need to use an iron filer or chlorination to remove the iron.

The hardness of your water and the amount to be softened will determine the size of your water softener. In terms of user-friendliness, a manual model requires the most assistance from a homeowner, and an automatic unit requires the least assistance.

The Least You Need to Know

➤ As a nation, we are dependent on clean, easily available ground water in both municipal water systems and individual wells.

➤ A private water system puts the onus of responsibility for its upkeep and safety directly on its owner.

➤ Ground water advocates and well owners are true believers and have raised the standards for water system contractors and their work practices.

➤ Yearly water testing is a must and should be supplemented by other testing whenever the water looks, tastes, or smells suspicious.

➤ Shock chlorination is a simple process that, when done properly, can really clean out a lot of harmful bacteria from your well water.

Glossary

ABS   Acrylonitrile butadiene styrene, a type of plastic used for drain lines.

access panel   An intentionally built opening that allows access to a fixture's pipes.

acrylic   A durable thermoplastic used to form bathtubs, tub/shower combinations, and jetted tubs.

adapter   A fitting that allows different types of pipe to be connected, such as galvanized steel and copper.

aerator   A small device that screws onto the end of a spout and mixes air with water for a smoother flow.

angle stop   Typical shutoff valve for a sink or lavatory.

anode rod   Often made from magnesium, the anode rod hangs inside a hot water tank and protects its lining from corrosion.

ball cock   A toilet's water intake valve.

bidet   A plumbing fixture that includes a faucet and sprayer for washing the user's underside while sitting.

cement backerboard   A building material, available in sheets, that is composed of fiberglass and cement and used as an underlayment for tile.

cleanout   A removable plug in a drain or waste pipe that provides access for clearing away any blockage.

closet auger   A snake used for clearing a clogged toilet.

closet bend   A fitting that connects a toilet's closet flange to the drain.

closet flange   The flange or hardware, attached to the floor, to which a toilet is bolted and secured.

code   A set of local or national rules that determine minimum safe building practices; specific codes, such as those for plumbing and electrical, control specific aspects of construction.

compression faucet   A faucet that works by compressing a washer to prevent water from flowing out the spout and then decompresses to allow water to flow.

copper pipe   A universally accepted pipe and one favored by many plumbers.

CPC   Chlorinated polyvinyl chloride, a type of plastic pipe.

Crapper, Thomas   An English plumber who did *not* invent the flush toilet, even though it is attributed to him.

cultured stone   Stone made up of ground-up pieces of real stone mixed with polyester resins and sealed with a protective gel coat.

dip tube   A pipe that replaces cold water inside a hot water heater and then empties near the bottom of the tank.

diverter   A valve that directs water to different outlets, such as from a tub spout to a shower head.

drywall   A term describing sheets of paper-coated, gypsum-based wall covering.

DWV   Abbreviation for drain-waste-vent system.

elbow   A pipe fitting with two openings that is used when the direction of the pipe must be changed, such as at a corner; also called an ell.

enamel, vitreous   Material applied to cast iron or steel fixtures for its durability and ease of cleaning.

escutcheon   A finish cover that conceals a faucet stem where it exits a wall or fixture.

fiberglass   A spun-glass product used to form bathtubs and tub/shower units.

fixture   A sink, tub, toilet, bidet, and any other such device that supplies water, possibly stores it, and allows for its disposal.

float arm/ball assembly   A toilet component that shuts off the water intake valve when a predetermined water level has been reached.

flux   An antioxidizing paste used when soldering copper pipe.

galvanized pipe   An older water supply pipe that is no longer used in residential plumbing; galvanized pipe is connected by threaded fittings.

GPM   Abbreviation for gallons per minute, a measurement of water flow.

gravity tank toilet   The most common toilet model that flushes away wastes when a volume of water is released by a tank.

hose bib   A shorter version of a faucet, used most often in a house to supply water to a washing machine.

hydrologic cycle   Also known as the water cycle, this is the evaporation of water and its return to the earth in some form of precipitation.

inspection   This is a requirement for many plumbing jobs; your local building department inspects your work to confirm that it meets code requirements.

lavatory   A fixed bowl or basin for washing.

low-flow toilets   Toilets that only use 1.6 gallons per flush.

main water supply pipe   The pipe that brings water into your house from the city main; also called a service line.

PB   Polybutylene, a flexible plastic tubing allowed by some codes as a water supply pipe; also the subject of class action law suits due to leaks.

permit   Permission from your local building department to proceed with certain plumbing or other construction work.

plumber's putty   Pliable putty used to seal joints between drain pieces and fixture surfaces.

plumber's tape   A perforated, flexible metal strapping used to secure pipe to framing lumber.

plunger   A tool for clearing clogged drains and toilets that consists of a rubber suction cup on the end of a wood handle; also called a plumber's friend.

pop-up assembly   One type of drain mechanism for lavatories and tubs.

porcelain   White ceramic ware applied to plumbing fixtures during a high-temperature firing.

potable   Clean, drinkable water.

pressure balance valve   A safety valve used for showers to maintain a relatively constant temperature as the water pressure changes.

pressurized toilet   A toilet with a separate tank that stores water. As the water enters, it compresses the air present in the tank, which then releases pressurized water into the bowl and out the trapway.

PVC   Polyvinyl chloride, a rigid plastic used in the DWV system.

reducer   A fitting that connects pipes of different sizes.

refill tube   Directs water into a toilet bowl after flushing.

riser   The vertical pipes in your system that carry water upward.

rough-in   The initial installation of water supply pipes and DWV pipes to their fixture locations, but before connecting with those fixtures.

shutoff   A valve that blocks the flow of water to a pipe or faucet.

sillcock   A valve that comes attached to a section of thicker pipe and used for outdoor use. Frost-free sillcocks are often required by local codes.

soil stack   Also called the main stack, this is the biggest vertical pipe in the DWV system; all the drain and waste lines empty into the soil stack, which both carries waste to the sewer line and vents the system through the roof.

solder   In plumbing, a metal alloy that, when heated, melts and then seals pipe fittings as they connect to sections of pipe.

stop valve   A fixture's individual shutoff valve.

tee   A T-shaped fitting with three openings.

temperature-pressure relief valve   A hot water heater safety device that prevents excess pressure from building up inside the tank by opening and releasing excessively hot water or steam out the discharge pipe.

trap   The curved section of a fixture's drain line that retains a small amount of water to prevent sewer gases from entering the house.

tube cutter   A tool made specifically for cutting copper tubing as well as plastic pipe.

union   Three-piece fitting, used primarily with steel water supply pipes, that joins two sections of pipe.

valve seat   The section of a compression faucet into which the stem fits.

valve-seat reamer   A tool for cleaning old, worn, and damaged valve seats in compression faucets.

vanity   A bathroom cabinet that also houses the lavatory.

vent stack   The top section of the soil stack that vents out through the roof.

vitreous china   A fired clay product with a low-porosity glass surface used to manufacture some toilets and lavatories.

washerless faucet   A faucet that uses different mechanisms, including cartridges, rotating balls, and ceramic discs, to control the flow of water without requiring compression.

waste and overflow   A bathtub's drain assembly.

water closet   An old name for a toilet, and one that's still used in the United Kingdom.

WYE   A Y-shaped fitting with three openings used to create branch lines.

Resources

This is the information age squared, and that applies to plumbing information as well. The following books and Web sites will further illuminate the wide world of pipes, drains, and fixtures.

Books

Better Homes and Gardens Step-By-Step Plumbing. Des Moines, Meredith Press, 1997.

Black & Decker The Complete Guide to Home Plumbing: A Comprehensive Manual, from Basic Repairs to Advanced Projects. Minnetonka, Minnesota, Cowles Creative Publishing, Inc. 1998.

Ogle, Maureen. *All the Modern Conveniences: American Household Plumbing, 1840–1890.* Johns Hopkins University Press, 1996.

Sunset Books Basic Plumbing. 4th edition. Menlo Park, Sunset Books, 1995.

Web Sites

www.plbg.com—Among other information, this site features links to plumbing manufacturers from A to Z.

www.PlumbingNet.com—According to the site's statistics, this is one of the most visited plumbing sites on the Internet.

www.PlumbingWeb.com—This site features links to contractors, suppliers, manufacturers, and plumbing organizations.

www.toiletology.com—You'll find more information on toilets here than you ever wanted to know.

www.theplumber.com—This site has general plumbing information.

www.terrylove.com—This is my technical editor's site and is also one of the most visited plumbing sites on the Internet.

www.leeps.com—This site is "The World's Largest Online Plumbing Catalog."

www.rotorooter.com—This is the first name you usually think of when it comes to cleaning big drain clogs.

www.copper.org—Copper advocates here will tell you their views on the benefits of copper pipe.

www.naphcc.org—Plumbing-Heating-Cooling Contractors Association.

www.hometime.com—A remodeling site based on the popular television series.

www.soundhome.com—Good information on home repair and remodeling.

www.kohlerco.com, www.americanstandard.com, www.moen.com—These are three sites of popular manufacturers of plumbing fixtures and faucets.

Other Sources of Information and Advice

EPA Drinking Water Hotline, 1-800-426-4791 (Washington, D.C.)—The EPA provides general information about federal drinking water regulations and guidelines.

Water Quality Association, 630-505-0160 (Illinois)—The WQA maintains a register of information on the effectiveness of commercially available water treatment equipment.

National Ground Water Association, 614-337-1949 (Ohio)—The NGWA is a 23,000-member international organization representing all professions of the ground water industry.

Center for Disease Control and Prevention, 404-639-2206 (Atlanta)—This is the federal center of expertise.

American Ground Water Trust
16 Centre St.
Concord, New Hampshire 03301
Phone: 603-228-5444
Fax: 603-228-6557
Web site: www.agwt.org

According to their Web site, "The Mission of the American Ground Water Trust is to: protect America's ground water, promote public awareness of the environmental and economic importance of ground water, provide accurate information to assist public participation in water resources decisions."

Index

CKBI (certified kitchen and bath-room installer), 295
CKD (certified kitchen designer), 295
clamp kit, pipes, leak repairs, 171
claw-foot cast-iron tubs, 115
clean-out plug, main water supply line, removing clogs, 135
cleaning
    after contractor work complete, 79
    disposers, 245
    faucet aerators, 192
    kitchen fixture surfaces, 118-119
    toilets, 158
clear coat sealant, 190
clevis strap (lavatory), 130
clogs, 126
    clearing out main water supply lines, 135
    disposers, 244
    kitchen sinks, 126-127
        augers, 129
        removing the trap, 127-129
    lavatories, 129
        plungers, 130-131
        removing stoppers, 130
        removing the trap, 131
    maintenance of drains after removal, 136
    tubs and showers, 131-134
        metal stopper with drain flange, 131
        pop-up drains, 133
        trip-lever drains, 134
closet augers, 86
codes, 28-29
    bathroom addition requirements, 278
    beta versions of products, 32
    history of the Uniform Plumbing Code, 30
    loop vent installations, 296
    national plumbing codes, 28-29
cold water
    branch lines, 9
    routes, upgrading, 53
coliform bacteria, 42
commodity tube copper pipes, 104
communication, hiring professionals, 74
composting toilets, 161-162
compounds, 4
compression
    faucets repairs, 56, 182-184
        damaged valve seats, 184-185
        replacing washers, 183-184
    fittings, 108
    valves, leaks, 178
computer appliances, remodeling kitchens, 293
condensation
    hot water heaters, 324
    toilets, repair, 146

conservation of water, 13
    measures, 13
    solar water heaters, 331
construction rules, bathroom additions, 285-286
contamination wells, 362-363
continuous chlorination wells, 366-367
continuous feed disposers, 241
contractors, 75
    allowing substitutions, 77
    changing orders, 78-79
    clean-up and wall repair, 79
    comparing bids, 78
    customer responsibilities, 79
    drilling (wells), 360
    general versus subcontractors, 257
    giving specifications, 77
    insurance, 77
    licensing, 76
    surety bonds, 76-77
    written contracts, 78
contracts, written, 78
control valves, installing sprinkler systems, 311
conversion of salt water to fresh water, 5
Cooperative State Research, Education, and Extension Service. See CSREES
copper pipes, 104
    annealed or soft-tempered, 104
    commodity tube or hard-tempered, 104
    cutting, 105
    joining, 106
        without solder, 107-108
    leaks, 166
    slip coupling, permanent leak repair, 175-176
    soldering, 106-107
    stop valve installation, 212
    unsoldering, 107
cordless drills, 91
    bits, 92
    chuck sizes, 91
costs
    bathroom additions, 278
    demand tankless hot water heaters, 328-329
    hiring professionals, 74
    hot water heaters, 321
    remodeling/upgrading, 64-65
counter lavatories, 57
    one-piece vanity tops, 57
    self-rimming, 57
    under-counter, 57
counters, remodeling, 252-253
CPVC (chlorinated polyvinyl chloride), 32
    pipes, 98, 265
    plastic supply tubes, 266
crescent wrenches, 86
cross-linked polyethylene. See PEX pipes
CSREES (Cooperative State Research, Education, and Extension Service), 367

cubic feet per minute. See CFM
cultured stone fixtures, 117
curb stops, 9, 73
curtain installation, showers, 235
customer responsibilities, contractors, 79
cutting
    copper pipes, 105
    notches, repiping, 269
    plastic pipe, 99-100
    tools, 89-90
        hacksaw, 90
        hole saws, 90
        keyhole saws, 90
        pipe reamers, 90
        propane torch, 90
        reamers, 90
        tube or pipe benders, 90
        tube-cutters, 90
cycle of water, 4
    chemical concoctions, 4
    desalinization, 5
    unnatural cycles, 5

D

Dark Ages, plumbing history, 17
debris, safety, 37
dedicated
    circuits, 62
    electrical circuits, 253
demand tankless hot water heaters, 326-327
    cost, 328-329
    gas versus electric, 327
    life expectancy, 329
    selection, 327-328
deposit dissolving, showers, 197-198
depth requirements, installing sprinkler systems, 310-311
desalinization (water), 5
design
    kitchen remodeling, 290-291
        certified designers, 295
    laundry rooms, 334-335
    lavatories, 113-114
digging, depth requirements, installing sprinkler systems, 310-311
dip tubes, hot water heaters, 323
dishwashers, 246
    energy savings, 247
    remodeling kitchens, 297
    repairing versus replacing, 246-247
disinfection wells, 362
dispensers, hot water, 247
disposers, 61, 241
    batch-feed, 241
    cleaning, 245
    continuous feed, 241
    installation, 221, 245-246
        drain pipes, 221
    problems, 242
        clogs, 244
        obstructed blade, 244
        overheated motor, 244

I